Disaster Management

Edited by

Harsh K Gupta

Universities Press

DISASTER MANAGEMENT

UNIVERSITIES PRESS (INDIA) PRIVATE LIMITED

Registered Office
3-6-747/1/A & 3-6-754/1, Himayatnagar
Hyderabad 500 029, Telangana, India
info@universitiespress.com; www.universitiespress.com

Distributed by
Orient Blackswan Private Limited

Registered Office
3-6-752 Himayatnagar, Hyderabad 500 029, Telangana, India

Other Offices
Bengaluru, Chennai, Guwahati, Hyderabad, Kolkata
Mumbai, New Delhi, Noida, Patna, Visakhapatnam

First published 2003
Reprinted 2004, 2007, 2010, 2011, 2013, 2017, 2019, 2021, 2023

ISBN: 978-81-7371-456-6

501874

Printed in India by
Yash Printographics Noida

Published by
Universities Press (India) Private Limited
3-6-747/1/A & 3-6-754/1, Himayatnagar
Hyderabad 500 029, Telangana, India

CONTENTS

FOREWORD

The vulnerability of the Indian subcontinent towards disasters, both natural and man-made, is widely recognized. Its unique subcontinental dimensions, coupled with facts like its geographical location and the behavior of the monsoon, make it one of the most hazard-prone countries in the world. India is vulnerable to various natural disasters like floods, droughts, cyclone, earthquakes, landslides, avalanches, forest fires, and the like. Disasters occur with great regularity and, despite preparedness to meet all contingencies, the economic and social costs on account of losses caused by natural disasters continue to mount year after year.

The Indian landmass covers over 30 lakhs sq km, with widely varying topography and terrain characteristics—coastal, mountainous and plain lands, each frequented by almost every type of natural hazard. Almost every year, one or the other water- and climate-related disasters, namely floods, droughts, cyclonic storms, landslides, snow avalanches, earthquakes, and coastal erosion occur in different seasons. These disasters cause widespread damage to life and property. By far, floods account for the biggest economic devastation in the country. Over 75% of the annual precipitation of 400 million hectare meters is received during the monsoon months, causing severe floods in several riverine regions of the country. The area liable to floods is estimated as 40 million hectares, with 8 million hectares being perennially affected. The average annual loss is of the order of Rs. 1340 crores. Through several flood control and protection measures, about 14.37 million hectares have been protected up to 1993. The droughts, though less frequent compared to the floods, have a greater potential for damage, engulfing large areas at any single time, and being of a prolonged nature. Due to the erratic behaviour of monsoon rainfall, over 68% of the Indian landmass is prone to drought incidence of varying degrees. While the severest droughts are found to occur once in 32 years, every third year experiences a drought of some intensity. Due to several large-scale programs, such as the National Watershed Development Program and the Integrated Watershed Development Program, several areas of the country are drought-proof to a large extent. The vast coastal belt of India is vulnerable to cyclonic storms, moving across the Bay of Bengal and the Arabian Sea. Tropical cyclones developing in the Bay of Bengal and the Arabian Seas affect the Indian coast with a frequency of 5 to 6 in a year, occurring both in the pre- and post-monsoon seasons. While the entire east coast is vulnerable to cyclones, on the west coast Gujarat and Maharashtra are mainly affected. The most recent example is that of the Orissa Super Cyclone of October 1999, which was the worst phenomenon in recent history. Hitting the Orissa coast at a speed of 250 kmph, it devastated a 250-km stretch of the coastal belt in eastern Orissa, killing around ten thousand people. The seashores are subjected to periodic arrival of cyclones and waves, with long-term wearing away of rocky cliffs and coastal erosion. Besides the above disasters, the high reaches of Himalaya are affected by avalanches during the snowfall season, causing loss of life and property. The states of Kashmir, Himachal Pradesh and Uttranchal are the most affected.

Earthquakes occur mostly along the seismically-active zones where different earth plates meet each other. In this context, it will be of interest to know that the Indian subcontinent was somewhere in the neighbourhood of Antarctica about 120 million years ago. Since then, the Indian Plate moved northwards and started pushing against the Eurasian plate about 40 million years ago. At the boundary, where the Indian plate pushes against the Eurasian plate, the Himalaya was created in five pulses. It will be of further interest to know that the Indian plate is still moving north-northeastwards at the rate of about 5 cm per year and, consequently, the Himalaya is still rising and the boundary areas are seismically very active—right from the Hindukush to Himachal Pradesh, to the Indo-Nepal border, to Assam. This inter-plate boundary is active seismically and great earthquakes have taken place, and will continue to take place, along the inter-plate boundary extending from the Hindukush to Terai to Assam. Unlike meteorology, the science of seismology has not yet reached a stage where earthquakes can be forecast. Research is on in this direction all over the world; our scientists are also active. The forecasting of earthquakes could be possible in the foreseeable future in certain geologically favorable locales.

The United Nation General Assembly, through a resolution, proclaimed the 1990s as the "International Decade of Natural Disaster Reduction" (IDNDR). The expectation of the IDNDR was to bring about measures that have a direct bearing on mitigation against natural disasters. Over the years, there has been some success in tackling these disasters. The international decade provided an opportunity to impart a sense of urgency to improve our strategies for reducing the impact of natural disasters and to ensure an integrated approach to achieve the overall objectives of mitigating the impact of natural disasters.

In the last year of the IDNDR, the INSA thought it appropriate to organize a one-day seminar on "Natural Hazards in India". In this seminar, the aforesaid natural hazards were addressed by eminent persons with wide-ranging experience in their respective field of activities; and the Academy is now bringing out the proceedings of the seminar, in the form of a special volume, to disseminate the information for the wider scientific planning and decision-making communities.

Harsh Gupta
Department of Ocean Development
Mahasagar Bhavan, C G O Complex
Lodhi Road, New Delhi

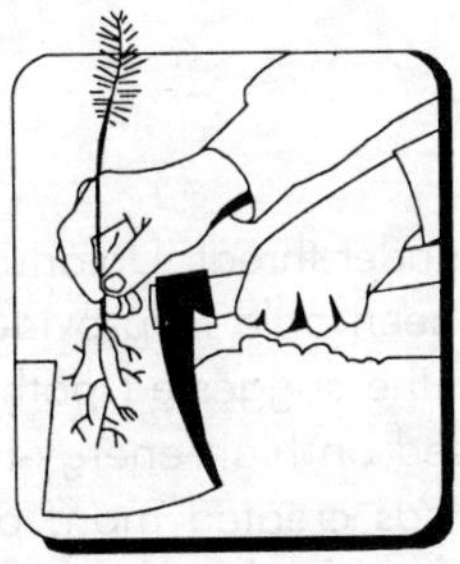

Vegetal Cover and Natural Disaster Management in Himalaya

P S Ramakrishnan*

ABSTRACT

The Himalaya, stretching over 2500 km from northeast to northwest, is a unique mountain system in that, within it are a variety of complex ecosystems in a biophysical sense, closely integrated into equally complex and diverse social systems. Being rich in forest, water and mineral resources, external pressure from outside the region have been active in exploiting them, for over a hundred years now. Large-scale deforestation is the consequence. The increase in population pressure from within the mountain region, largely exacerbated by external pressures from the industrial societies from the plains, has contributed to major changes in the land use patterns in these mountain ranges and the associated rapid depletion of natural resources. The intrinsically fragile geology of these relatively younger mountain formations are subject to constant stress under the influence of increasing anthropogenic influences. Therefore, the whole issue of disaster management for the Himalayan region need to be viewed in the context of the rapid land use and land cover changes. The mountain societies whose agriculture and other economic activities are closely interlinked with forest resources (Ramakrishnan *et al.*, 1994a, 1996a,b) have been forced to operate under an ecological situation of depleted forest resource base. The net result has been large-scale transformations in the land use, forest plantations raised largely for industrial purposes by the foresters, degraded landscapes often with biological invasion (Ramakrishnan, 1991) and, in extreme cases, site desertification, where the land loses all its productivity. Since a large section of the mountain societies are still largely dependent upon land-based activities,

* School of Environmental Sciences, Jawaharlal Nehru University, New Delhi 110 006

their livelihood concerns are under threat. Unfortunately, the textbook-based, scientific knowledge has not been able to provide a viable alternative land use developmental pathway; the suggested pathway for a major land use activity like agriculture is based on high energy-input, terraced agricultural system, which by and large has created more problems than it has been able to solve. The issues involved for land use management are more than biophysical; social, economical cultural and institutional concerns have often been missed out. It is in this context that the following discussion becomes significant. From the disaster management point of view, the issues involved are coping with the ecological fragility of the mountains, one on the one hand, and dealing with the general apathy of the mountain societies towards the pathways adopted for development, which, on the other, are of no consequence to them.

Land Use System Dynamics

Land use and cover change in the mountain environment of the country is significant to a whole range of themes and issues of a biogeochemical nature, central to the study of 'global change', which encompasses within it a whole variety of such concerns as global warming and climate change, biodiversity depletion, biological invasion, land degradation, and desertification (Ramakrishnan *et al.*, 1996b, 1999). By implication, land use dynamics is of focal concern for the sustainable development of the natural resource-rich mountain regions (Ramakrishnan, 1992. Ramakrishnan *et al.*, 1994a). On a local/regional scale, by altering ecosystem complexity in a variety of different ways – ranging from the subspecific, through species, ecosystems and landscape levels – land use and cover change could impact upon the sustainable livelihood concerns of local communities and alter the ecosystem fragility. On a global scale, it would have implications for changes occurring in the biogeochemical cycles of the earth, and atmospheric levels of greenhouse and other trace gases. Here, we are concerned with the local/regional issues, ecosystem fragility, and disaster management and the linked sustainable management of the landscape and livelihood concerns of mountain societies, in the context of ecological, socio-economic and cultural dimensions of the problem.

Traditional Mountain Societies and Ecosystem Complexity

Traditional mountain societies are characterized by their close interconnection that they maintain with nature and the natural resources that it contains. They depend upon the natural resources, in general, and the biodiversity (ecosystem complexity ranging from the subspecific through species, ecosystem, and landscape levels; cf. Ramakrishnan, 1996) contained therein in particular, as the chief driver for their sustainable livelihood concerns (Ramakrishnan, 1992; Ramakrishnan *et al.*, 1994a, 1996b). This relationship extends beyond the economic realm; social, cultural, and spiritual dimensions also play a significant role in this linkage with nature and natural resources (Ramakrishnan *et al.*, 1998). They view the ecosystem and the social system in which they operate as a unified whole. Arising out of this relationship with nature, based on coexistence rather than competition, the sustainable use of natural resources is an adaptation, based on a two-way interactive process. The net result is a set of institutional arrangements evolved towards ecological prudence, the ultimate objective being sustainable use of natural resources to cope with uncertainties in the environment, rather than a short-term strategy to maximize production. The traditional ecological knowledge, centred around manipulation of biodiversity, determines the land use dynamics. The end result is a land use system that is based on diversification rather than homogenization of the landscape. With a variety of incursions made by modern society, in recent times, their linkages with nature has often been adversely altered, resulting in social disruption (Ramakrishnan, 1996a,b). Yet, the efforts to impose a technological quick-fix for their economic development has often been unsuccessful, because these are too often based upon a value system these societies cannot understand or appreciate, and therefore participate in the developmental process itself (Ramakrishnan *et al.*, 1994a).

Ecosystem Complexity in the Context of Traditional Wisdom

Many traditional mountain societies, all over the world, owing to their close linkage with nature and natural resources and because of animistic belief systems, have protected either an entire mountain system or refugia of their natural ecosystem/s as 'sacred' (Ramakrishnan, 1996; Ramakrishnan *et al.*, 1998). The Himalayan societies also have a rich traditional ecological knowledge base, which is embedded both within and outside the realm of their cultural ecology. Their traditional culture manifests itself through natural resource-linked concepts such as sacred species, groves and landscapes—the guiding principles that regulate the use of natural resources being embedded in their socioculture-based institutions.

Here, we argue that emergence of these institutions were intended to boost social solidarity through religious norms, rather than promoting environmental consciousness *per se*. The specificities of these institutions are codified with particular religious myths and beliefs. The demarcation of the sacred areas, in the past, from the rest in a natural resource-rich background of the 'traditional societies' strengthens the above argument. However, sociocultural practices and traditional ecological knowledge embedded within the realm of these beliefs have important implications for conservation of the natural resources in the contemporary context of often highly-degraded, natural resource-poor mountain slopes; the ecological issues involved are discussed here.

The Concept of Sacred Groves

Sacred groves are part of a landscape, often a forested ecosystem, with well defined geographical features, delimited and protected by traditional societies for cultural/religious reasons (Ramakrishnan *et al.*, 1998). Often associated with a mountain landscape, these groves are found in many parts of the world (Hughes and Chandran, 1998). However, these groves, which were once widespread, have disappeared more or less completely from certain parts of the world like Europe, and are at varied levels of protection in other areas where they exist. On the other hand, planted trees that are often sacred and considered as sacred groves, and found in many religious temples of the Asian region, are a more recent development.

Since many forest-dwelling, traditional societies invariably have sacred groves, it is reasonable to conclude that this remains the more ancient in the social evolutionary scale, where a village or a cluster of villages that maintain them have their own cultural norms and limited institutional arrangements governing their protection. These institutional mechanisms, often not codified, have helped in protecting the rich biodiversity contained therein, through a whole set of myths and beliefs; protection was reinforced by recognizing them as the focal point for varied levels of social interaction.

Divergent viewpoints exist on the origin of this concept amongst traditional societies—arguments ranging from the fear and respect for God and ancestral spirits, to more secular causes such as the utilitarian values of biodiversity contained therein. In any case, it is indeed true that these groves have functioned in the past, and still continue to function, as sites for socio-cultural and religious interactions. Obviously then, there is need for a variety of institutional arrangements, and if they do exist, to effectively maintain and manage these sacred sites. It is indeed true, in the contemporary context, that these sacred sites have often been an island of biodiversity in a degraded landscape created through a whole variety of human exploitation of natural resources; obviously, this cannot be a reasoning to suggest that traditional societies recognized the value of biodiversity when they were, in any case, part of a resource-rich environment.

From a contemporary point of view, the relict vegetation in an otherwise degraded landscape is important as a source for the germplasm required for a meaningful rehabilitation work (Ramakrishnan and Ram, 1988; Ram and Ramakrishnan, 1992; Kiewtam and Ramakrishnan, 1993).

The Concept of the Sacred Landscape

Due to their awe-inspiring features, mountains have been revered and have been central to the cultural ethos and religious belief systems of many societies all over the world. Ranging from a restricted landscape to the extent of covering the whole of a given mountain range, these sacred sites are valued for a whole variety of reasons.

The sacred landscape evolved from the level of protected segments (sacred ecosystems/groves) of a landscape, operates at two levels of hierarchy. The higher level has the least specificity linked to institutions, but has the greatest zone of influence. Least specificity means a lower number of prescriptions and prohibitions. An example of this is the conceptual sacred landscape traced by the Ganga river system (Table 1). Social institutions here are in a diffused state, though the sphere of influence is spread over the entire Indian subcontinent.

Table 1 *The Ganga mega-watershed: An example of a loosely-organized sacred landscape*

- One of the best examples linking highland/lowland interactions is represented by the course of Ganga river system in India, originating at Goumukh in the higher reaches of the Garhwal Himalaya, tracing through the northern plains of the States of Uttar Pradesh, Bihar and West Bengal, before the river drains into the Bay of Bengal in the east. The sacred land, the river tributaries, the human habitation, all the natural and human-managed ecosystems, a chain of temples dating back to antiquity, the holy cities of Gangotri, Jamnotri, Kedarnath, Badrinath, Rishikesh, Haridwar, etc, of the mountain zone, and those in the plains such as Allahabad and Varanasi, all together represent a set of interconnected ecosystems bound together by the sacred river itself.
- Worshipped by the Hindus and Buddhists of the Asian region and tucked away in the folds of the Himalaya, the symmetrical Mount Kailas rises above the Tibetan plateau, and is the legendary Mount Meru or Sumeru, the 'Mandala' of the Buddhists (the cosmic axis around which the axis of the Universe is

Continues...

Table 1 *Continued...*

organized for both Buddhists and the Hindus). This mythological interconnectedness belief system has even penetrated into the belief system of the distant Balinese of the Indonesian island complex. As the origin of all the major sacred rivers of Hindu mythology, and these river systems being the basis for the human civilization in this part of the world, it indeed plays a pivotal role in this part of the world; the conservation of natural resources linked to the Himalayan mountain region could be linked to these belief and value systems of the people.

Next in the hierarchy is the example of the 'Demajong' landscape in West Sikkim, based on Tibetan-Buddhist philosophy, with clearly defined norms, and a well-defined boundary for sacredness. The air, soil, water, and biota are all sacred; any perturbation in the landscape being restricted use being determined by the social institutions. With a variety of rituals linked to the diverse communities living within the landscape boundary, who have their own predetermined rights for natural resource use, larger community participation is ensured.

Small- versus Large-scale Perturbations

For a very long time during the course of the development of the present-day ecological paradigm, ecologists viewed ecosystems to be producer-controlled, both structurally and functionally, and hence predictable. The research emphasis was on well-protected, pristine ecosystems. During the 1970s, the emphasis shifted somewhat with the conceptualization of predator-controlled or consumer-controlled ecosystems. Still, the emphasis remained on undisturbed study sites. With greater realization and emphasis placed on evolution as an ongoing active process, in more recent times, disturbance, whether intrinsic or extrinsic to the system, was seen as a driving force in the evolution of ecological systems. Viewed in this context, the basis on which non-codified institutional arrangements of the 'Demajong landscape', as perceived by the Tibetan Buddhists, make interesting reading. Whilst small, human-made perturbations such as a variety of agricultural activities, fishing in the freshwater bodies, or other forest-linked economic activities are permissible for the local population living in the landscape area, large-scale perturbations are strictly prohibited. The myths and beliefs associated with this subtle distinction between small versus large perturbations is centred around the belief system that the hidden treasures (the 'ters'), both on the land and in the water-bodies, will be lost due to large perturbations. It was not surprising, therefore, that the Tibetan Buddhists reacted strongly when there was an attempt by the Governmental agencies to have a hydroelectric project in the region, through the diversion of the 'Rothang Chu' River and the desecration of the land and water bodies arising therefrom. This is the context in which the large-scale deforestation and conversions of land use systems have to be viewed, in the context of ecosystem fragility and its implications for disaster management.

The concept of the sacred species

Within both the sacred grove and the landscape are often found sacred species, which are socially, culturally or religiously valued (the oaks locally called 'Bhanj' in the Central Himalayan Kumaon and Garhwal, discussed in the context of forest management). Unlike in the case of the other two categories, the sacred grove and the landscape, a utilitarian viewpoint could be given to the concept of the sacred species, since many of them have ecological or socioeconomic values as we now understand them. Being a direct product of traditional ecological knowledge accumulated by the human civilization, on the basis of experiences, this is a more recent product in the social evolutionary scale (Ramakrishnan *et al.*, 1998). We will be discussing them only as components within the other two sacred categories (though it exists even outside them), for their value for biodiversity management and, in turn, for disaster control.

Sustainable Landscape Management

Both within and outside the sacred landscape, in the biodiversity-rich areas of the tropics, exist a large number of sociocultural societal entities linked to nature and the natural resources contained therein. These linkages extend through their traditional agricultural practices, fuelwood and timber extraction from the forests, dependence on non-timber forest products, etc. The traditional societies often also perceive other not-so obvious benefits through their natural resources—such as the role of agricultural and forest biodiversity in soil fertility maintenance and for establishing synchrony in nutrient availability, both in space and time (Ramakrishnan, 1992), and in ensuring water balance in the soil profile through ecosystem management strategies (e.g., the culturally-valued oak species/forests in the Central Himalaya are viewed as helpful in maintaining water balance in the soil profile) (Ramakrishnan *et al.*, 1988). Therefore, the way in which ecosystems are managed within the landscape becomes important for ecosystem/landscape integrity. We shall briefly look at some of the issues involved in sustainable management of ecosystem/landscape, with community participation.

Traditional agroecosystems and their management

The large variety of multi-species agroecosystems maintained by traditional societies (Figure 1), derive a variety of inputs from the landscape

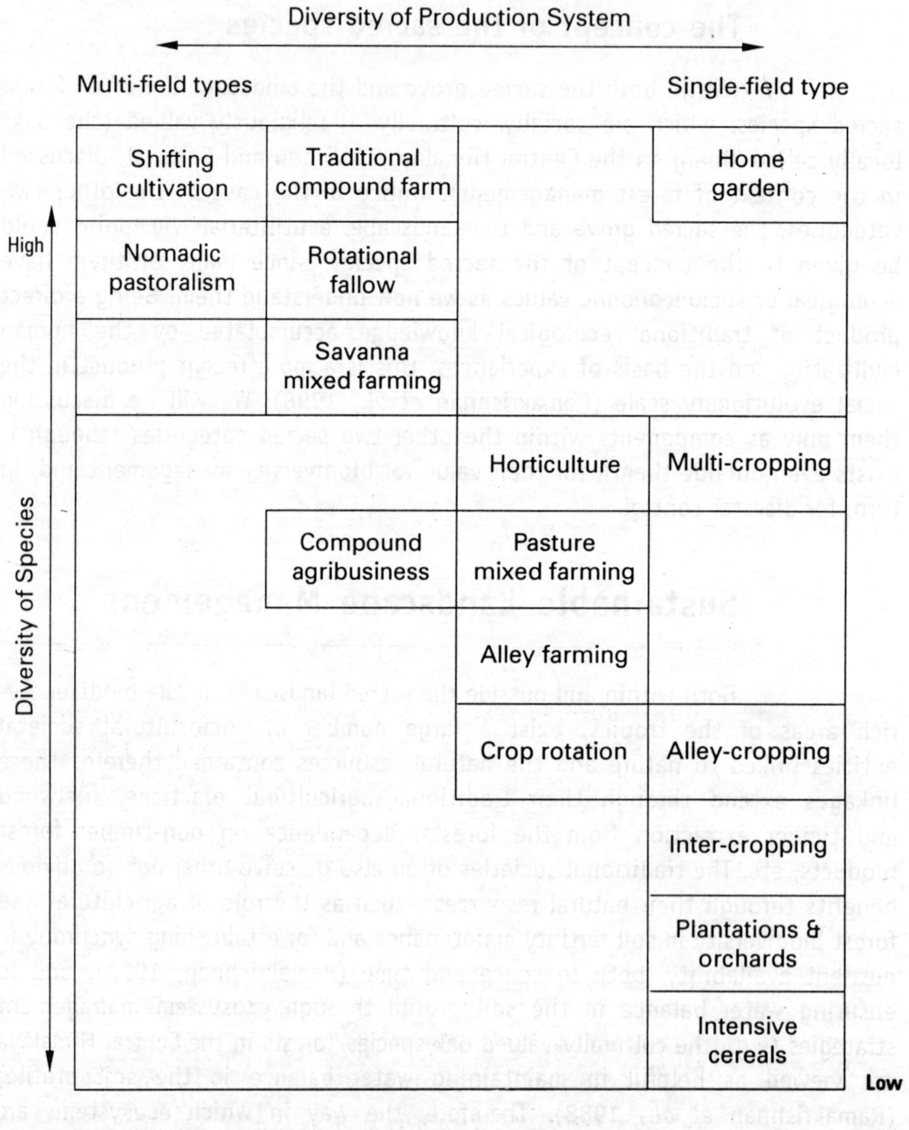

Figure 1 *Broad typology of agricultural systems that could exist in a sacred mountain landscape, organized on the basis of their biological diversity (from Swift and Ingram, 1996).*

system in which they operate. Ranging from shifting agriculture, which is an agroforestry system separated in time, a whole range of agroforestry systems organized in space represent an attempt on the part of the mountain farmer to 'imitate' the way the natural ecosystem/landscape functions. It is, therefore, not surprising to find recycling of resource between agriculture and forest systems as an important functions in the set of process-based linkages (Ramakrishnan, 1994; Swift *et al.*, 1996; Ramakrishnan *et al.*, 1993, 1999).

This biodiversity contributes in a variety of ways towards ecosystem functions such as production, decomposition, nutrient-cycling dynamics and thus towards stability and resilience of the system, with varied levels of management ranging from the casual to high intensity, eventually leading to modern monocropping systems (Figure 1) (Swift, *et al.*, 1996). This is qualitatively different from the 'modern agriculture' which essentially represents a monoculture of one crop species/cultivar, a highly simplified landscape, supported by the whole set of energy subsidies, rather than operating as an integral component of a self-sustaining complex landscape unit, mediated through humans. The well-conceived integration of complexities with a number of ecosystem types as units within the given landscape, therefore, becomes significant.

The cultivation of multiple species by farmers necessarily entails biodiversity issues. Farmers' choices of cropping combinations represent a planned biodiversity strategy which, together with crop and management practices, impact upon other organisms causing changes in population structures and sizes in what may be termed the associated biodiversity of the system. These structural changes, in turn, influence production. Thus, the shifting agriculture farmers in Northeast India (Ramakrishnan, 1992a) have a variety of cropping systems related to the shortening of the agricultural cycle and the related site quality characteristics, leading to rotational fallows and sedentary systems (Figure 2). Indeed, even on the same site on a hill-slope, crop organization is to a large extent determined by the nutrient distribution pattern; with emphasis on nutrient use-efficient tuber and vegetable crops on the top of the hill-slope and less efficient crops emphasized in the nutrient-rich bottom of the slope. Emphasizing upon weed management rather than weed removal, the Northeast Indian farmers, like their counterparts from the distant Mayan region of southern Mexico (Gliessman, 1988), manipulate overall biodiversity, in turn influencing various ecosystem functions including the processes linked with resource capture by the crops.

That multi-species land use systems are more stable, more productive, and less risky, compared to fewer species under conditions of 'global change', could be conjectured from the fact that greater diversity in above- and below-ground structures results in increased ability to capture resources as in mixed cropping,

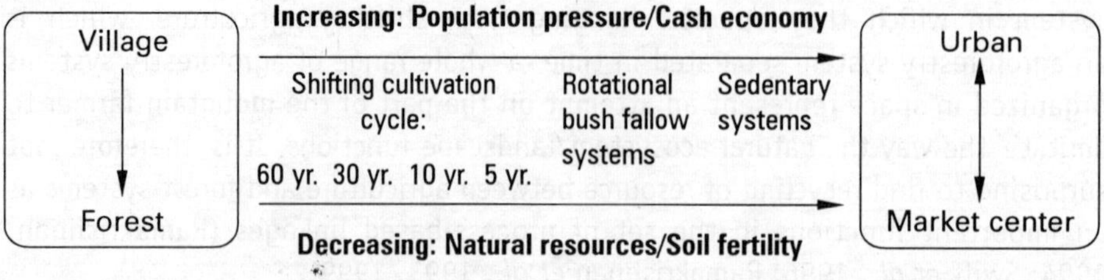

Figure 2 *The land use changes as related to population pressure, land degradation, and available linkages to a market economy. These pressures decline as one moves away from the city center in the Northeast Indian uplands. Please note the adaptation of agricultural systems by traditional societies, from shifting cultivation to sedentary systems, brought about to cope with ecological and socioeconomic uncertainties (adapted from Ramakrishnan, 1992a).*

shifting agriculture, and traditional home garden systems (Gliessman, 1990; Ramakrishnan, 1992a, 1994). Mixed cropping systems are also receiving attention from modern agricultural scientists for their role in biological pest suppression (Litsinger and Moody, 1976; Letourneau, 1990; Altieri, *et al.*, 1990). An important result that could accrue from a greater understanding of the relationship between indirect biodiversity and ecosystem function is to widen the scope of planned biodiversity, in effect, strengthening the management capabilities within agroecosystems.

It is generally acknowledged that biodiversity decreases as habitats change from forest to traditional agricultural systems, and then onto modern agriculture (Gliessman, 1990; Ramakrishnan, 1992a; Swift *et al.*, 1996) as a function of increasing intensity of management. However, biodiversity concerns so far have largely been confined to natural ecosystem management and the consequences of conversion of a forest to agriculture, while different agricultural systems themselves are not adequately emphasized. While a variety of models for loss in biodiversity under varied intensities of management regimes for agriculture are proposed, it seems to follow from recent studies that the biodiversity decline is sharp—somewhere in the area close to the middle intensity of management (Swift, *et al.*, 1996). If that be so, it is crucial to have a level of management that is closer to this critical area for sustaining biodiversity in agriculture, at the same time increasing productivity through appropriate scientific inputs.

If we consider high-input modern agriculture as only one of the possible pathways for agricultural development, one could have at least two more additional pathways for sustainable agriculture: (a) evolution by *incremental change,* and (b) restoration through the *contour pathway*. These two pathways differ from modern agriculture, which is an artificial entity standing apart form the rest of the landscape—an attempt to convert the natural ecosystem into one that contains only those biological and chemical elements that the planner desires, almost irrespective of the background ecological conditions. (Swift, *et al.*, 1994). The *contour pathway,* on the other hand, seeks to acknowledge and work with the ecological forces that provide the base on which the system must be built, well integrated into the landscape unit, while acknowledging at the same time the social, economic, and cultural requirements of the farming communities. Working with nature, rather than dominating it, this approach would involve active planning with the nature of the background ecosystem being kept fully in mind. Many agroforestry system types in the 'low' and 'middle' intensity management categories (cf. Figure 1) will come under this pathway. The *incremental pathway,* also is aimed towards ensuring landscape integrity. Many traditional agricultural systems need to be redeveloped through incremental, rather than quantum change, based on traditional ecological knowledge; anything drastic may not find acceptance by the local communities. In this incremental change towards sustainable development, one may have to consider a short-term strategy that may be constrained because of ecological, economic, social or cultural reasons, apart from a more ideal and perhaps desirable long-term strategy. One of the good examples of the *incremental pathway,* is the Northeast Indian case study in the shifting agricultural country, where models are built towards strengthening the forestry component of shifting agriculture that has gone weak for that selfsame reason; the Nepalese Alder (*Alnus nepalensis*) is one such example of an ecologically significant keystone species, which contributes to soil fertility through nitrogen fixation and is also a socially-valued species by the tribal societies in Northeast India. (Ramakrishnan, 1992a). The conclusions arising out of this analysis, which is now being implemented in over 1200 villages in Nagaland in Northeast India (Faminow, 1999), has indeed wider applications for this land use system prevalent all over Asia, Africa and Latin America (Table 2).

Table 2 *Shifting agriculture "jhum" and sustainable development for Northeastern India (from Ramakrishnan, 1992a)*

- For improving the system of land use and resource management in Northeastern India, the following strategies suggested by Ramakrishnan and his coworkers are based on a multidisciplinary analysis. Many of these proposals have already been put into practice.
- With wide variations in cropping and yield patterns under *jhum* practiced by over a hundred tribes under diverse ecological situations, where the transfer of technology from one tribe/area to another alone could improve the *jhum,*

Continues...

Table 2 *Continued...*

valley land and home garden ecosystems. Thus, for example, emphasis on potato at higher elevations compared to rice at lower elevations has led to a manifold increase in economic yield, despite low fertility of the more acidic soils at higher elevations.

- Maintaining a *jhum* cycle of a minimum of 10 years (this cycle length was found critical for sustainability), when *jhum* was evaluated using money, energy, soil fertility, biomass productivity, biodiversity, and water quality as currencies by greater emphasis on other land use systems such as the traditional valley cultivation or home gardens.
- Where the *jhum* cycle length cannot be increased beyond five-year period that is prevalent in the region, redesign and strengthen this agroforestry system incorporating ecological insights on tree architecture (e.g., the canopy form of tree should be compatible with the crop species at ground level, so as to permit sufficient light penetration and provide rapid recycling of nutrients through rapid leaf turnover rates). Local perceptions are extremely important in tree selection, for introduction into the cropping and fallow phases of *jhum,* as is being done in a major initiative in the state of Nagaland in Northeast India.
- Improve the nitrogen economy of *jhum* at the cropping and fallow phases by the introduction of nitrogen-fixing legumes and non-legumes. A species such as the Nepalese Alder (*Alnus nepalensis*) is readily adopted because it is based on the principle of adaptation of traditional knowledge to meet modern needs. Another such example is the lesser-known food crop legume *Flemingia vestita,* traditionally used by tribals as an important species when *jhum* cycles decline below 5 years.
- Some of the important bamboo species, highly valued by the tribals, can concentrate and conserve important nutrient elements such as N, P and K. They could also be used as windbreaks to check wind-blow loss of ash and nutrient losses in water.
- Speed up fallow regeneration after *jhum* by introducing fast-growing native shrubs and trees.
- Condense the time-span of forest succession and accelerate the restoration of degraded lands, based on an understanding of tree growth strategies and architecture, by adjusting the species mix in time and space.
- Improve animal husbandry through improved breeds of swine and poultry.
- Redevelop village ecosystems through the introduction of appropriate technology to relieve drudgery and improve energy efficiency (cooking stoves, agricultural implements, biogas generation, small hydroelectric projects, etc). Promote crafts such as smithying and products based on leather, bamboo, and other woods.

Continues...

Table 2 *Continued...*

- Strengthen conservation measures based upon the traditional knowledge and value system with which the tribal communities could identify; e.g., the revival of the sacred grove concept based on cultural tradition, which enabled each village to have a protected forest once upon a time, although few are now left.
- In the ultimate analysis, have an integrated approach for land use development in a given ecological/cultural landscape; base short-term sustainable livelihood strategy building upon traditional knowledge and technology; long-term sustainable development plans, based on larger ecologic/economic considerations, should gradually built up so as to avoid social disruption.

Forest Ecosystem Management

It is being increasingly realized that understanding the linkage between the ecological and social processes is the basis for designing ecological strategies for sustainable forestry management in the developing tropics (Ramakrishnan, 1992b). Ecological knowledge from such areas as tree biology and architecture (Ramakrishnan *et al.*, 1982), patch dynamics, ecophysiology of developing forest communities, reproductive biology, and nutrient cycling processes (Ramakrishnan, 1992a) could all be integrated into the current management process and future management options. In such an integrated approach to management, the socioeconomic and sociocultural issues and traditional knowledge coming from local communities need to be reconciled (Ramakrishnan, *et al.*, 1996b, 1998).

An improved quality of life to the local communities through forestry-linked agriculture (Ramakrishnan, 1992a), providing for the collection and usage of forestry linked, non-timber forest products, taking into account the societal perceptions of the quality of environment (Ramakrishnan *et al.*, 1999), and capitalizing upon the cultural perceptions of forests (Ramakrishnan *et al.*, 1998), also discussed above, are some of the considerations that should be taken on board, apart from sylvicultural considerations so far emphasized by foresters.

Among the many options that are available, large-scale agroforestry for the rural poor seems to be one. This option is attractive in that it combines food production and carbon sequestration through the forestry component, within the same system. Many of the multi-species, complex agroecosystems – being agroforestry systems of one form or the other – having fallen to disrepute on the face of onslaught from modern agriculture, there is a growing realization for redeveloping these systems on the lines discussed above. Such an option is becoming more and more popular in Asia and Africa, as a means for agricultural and rural development (Bentley *et al.*, 1993). In conjunction with social forestry activities, it forms a powerful tool for meeting a variety of rural needs such as food, fodder, fuelwood, timber and other forest products, and often also to meet not easily recognizable ecological benefits.

To cite one such example on the cultural dimension of forest management, one such set of species in the Garhwal region are *Quercus spp*. Associated with these species exist a lot of folk music, dance, literature and poetry; these species form good fodder, fuelwood and timber species, apart from the fact that they play a key role in soil fertility maintenance in human-managed ecosystems, nutrient cycling and conservation in natural forests, ensuring water balance within the soil, and thereby supporting biodiversity within the forest ecosystem (Ramakrishnan *et al.*, 1998). No wonder then, that the large-scale conversion of mixed oak forests to pine plantations by foresters over the last hundred years, in mid-elevation ranges of the Central Himalaya, formed a major causative factor for the now famous 'Chipko' ('hugging the tree') movement (Ramakrishnan *et al.*, 1998); the immediate cause being a reaction against timber extraction by forest contractors. The message that comes across loud and clear is that, these species that are socially valued by the community could play a key role not only in forest management but also in the rehabilitation of degraded systems, with peoples' participation.

Deforestation and forestry in the tropics is essentially a human problem; a narrow-based forestry initiative alone cannot tackle it. A broader-based multi-disciplinary approach will facilitate the sharing of knowledge arising out of the three major components of the management strategy (Figure 3).

Landscape Management

Realizing that biodiversity does contribute in a variety of ways to ecosystem functions (Gliessman, 1990; Ramakrishnan, 1992a) and that agroecosystems do harbour a great deal of biodiversity valuable for human welfare (Pimental *et al.*, 1992), it is reasonable that we go in for a mosaic of natural ecosystems coexisting with a wide variety of agroecosystem models derived through all the three pathways. Such a highly diversified landscape unit is likely to have a wide range of ecological niches conducive to enhancing biodiversity and, at the same time, ensuring sustainability of the managed landscape itself.

Traditionally, many societies have viewed their land use activity in a given landscape as part of an integrated land use management, wherein human managed ecosystems are closely linked to a variety of natural systems. (Ramakrishnan, 1992a; Ramakrishnan *et al.*, 1998). The diversity of cropping and

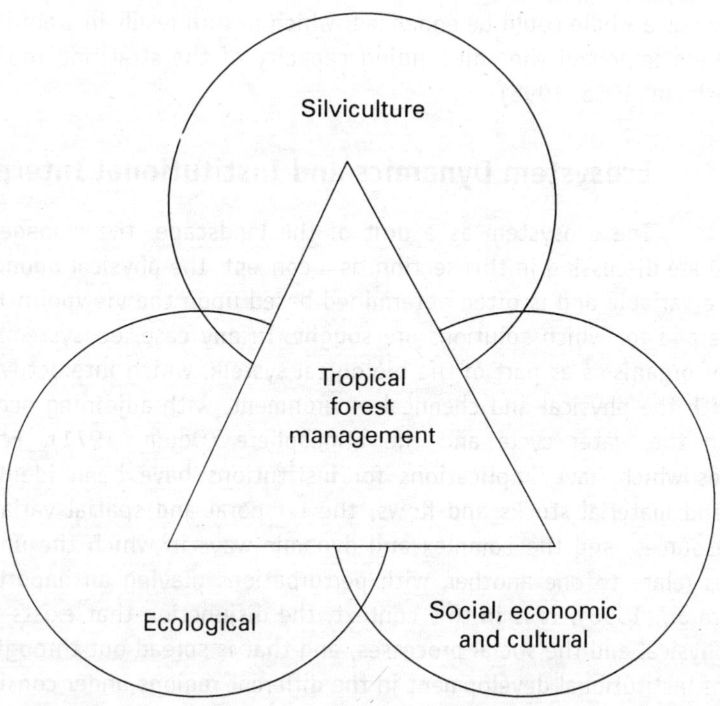

Figure 3 *Interdisciplinary interactions called for in tropical forest management and conservation (from Ramakrishnan, 1992b).*

resource systems that form part of the landscape, with a whole variety of coexisting tribal societies in Northeast India or that operates under similar situations elsewhere in the mountainous terrain, serve not only as a major means of protecting ecological integrity at the landscape level, but also acts as the knowledge and resource base that makes adaptivity possible (Maikhuri and Ramakrishnan, 1990; Ramakrishnan, 1992a; Brookfield and Padoch, 1994). Thus, the rich home gardens traditionally maintained by the Tara'n Dayaks of western Kalimantan of Indonesian Borneo, and currently being emphasized by this tribe along with raising irrigated rice in river valleys and along the base of the hill slopes by digging the higher ground to an irrigable level, is viewed by the authors as a response to population increase and new market opportunities. Such an adaptability is feasible from conserved diversity readily available at the landscape level.

The concept of protected 'sacred groves' and even the more complex 'sacred landscape' units are indicative of the approaches that contribute to landscape level complexity. At the species level, the role of the culturally valued oaks in the Central Himalayan region, discussed above, is an example how forest biodiversity with diversity in root and shoot architectural designs at the vegetation as a whole could be enhanced, which in turn result in stability of the soil through improved the soil-binding capacity of the stratified root system (Ramakrishnan, 1986, 1998).

Ecosystem Dynamics and Institutional Interplay

The ecosystem as a unit of the landscape, the management of which we are discussing in this section, is a concept, the physical boundaries of which are variable and is often determined based upon the viewpoint that one may take and for which solutions are sought. In any case, ecosystems have a variety of organisms as part of the biological system, which interact with each other, with the physical and chemical environment, with adjoining ecosystems and with the water cycle and the atmosphere (Odum, 1971). Ecosystem properties which have implications for institutions have been identified as energy and material stocks and flows, the temporal and spatial variability of those resources, and the complex and dynamic ways in which the underlying processes relate to one another, with perturbations playing an important role (Pritchard Jr., 1998). It is in this context, the disjunction that exists between the biophysical and the social processes, and that is spread out throughout the history of institutional development in the different regions under consideration here, has to be viewed. Indeed, an understanding of institutional interplay from a historical perspective should provide clues as to the ways in which land use-related institutions could be reorganized. Let us briefly look at the two different situations about which we have good information.

The Himalayan Blunder

The history of the development of present institutional arrangements in the Nanda Devi buffer zone area, for example, is illustrative of the *ad hoc* manner in which they were created without due consideration for natural resource management. Traditionally, the land and resource use-related institutional arrangements were informal and not codified, when resources were unlimited, and vested interest in them were perhaps minimal. However, in 1865, when the forest land (uncultivated land, including forests, pastures, and snow-clad peaks) were declared as the colonial government's public property, local people were allowed traditional forest use rights; simultaneously, private land use rights were also granted for cultivation. However, with more restrictions imposed on the use of the NTFPs from the forest in 1920 on forest resource use through the concept 'reserve forest', the government at that time introduced the concept of 'civil forest', assigned closer to their settlements, for use by local communities, but with land rights vested in the revenue department of the government; these forests were of poorer quality. With the recognition of the area as a wildlife sanctuary in 1939, more restrictions were imposed on reserve forest use, though opening up of the area for tourism became a new source of income to the people. The governmental response to the increasing resentment against restrictions was the creation of another institution called the 'community forests' locally called the 'panchayat forests', the ownership resting with the revenue department but the managerial rights with local communities. When large-scale timber extraction for industrial use from the buffer zone area started causing land degradation, the villagers responded by hugging the trees to protect them against felling, the now famous 'Chipko' movement.

The conversion of the Nanda Wildlife Sanctuary in 1982, and its further elevation to a biosphere reserve in 1988, and its recognition as a world heritage site in 1992, all led to more restrictions on forest use rights, and completed the exclusion of tourism. The institutional arrangements underwent drastic changes over a period of time, more as *ad hoc* arrangements to calm down the rising disenchantment of the local communities with the decision-making process, rather than that based on a logical framework determined by understanding local traditions and knowledge system linked to natural resource use. Indeed, this is a typical way in which natural resource management decisions are still being taken not only here but elsewhere, and indeed in many other parts of the world. Obviously, institutional arrangements have to be arrived at, on the basis of a good understanding of the ecological systems and the social systems that one is dealing with; where, of necessity, the existing confused arrangements arrived at over a period time and space need to be revamped by a systems overhaul. A hierarchy of systems development could be visualized where, local level institutional arrangements are revamped with a view to harmonize ecological systems with social systems, whilst the higher levels could be contributing to more generalized policy dispensations.

Conclusions

Disaster management in the mountains is a complex problem. The fragility of an uneven mountain landscape in which mountain societies exist gets compounded by geological factors as in a young mountain formation such as the Himalaya, and poorly-conceived infrastructural development activities. Disaster management strategy under such a situation has to be based on a multi-pronged approach. Technological solutions such as road alignment, urban planning, etc., have to be linked with ecological solutions, linked in turn, to the sociocultural perceptions of local communities (Ramakrishnan, 1998), based on ensuring complexity at ecosystem/landscape levels.

The more recently evolved 'Biosphere Reserve' concept of UNESCO, indeed a rediscovery of the sacred landscape concept, is an attempt towards such an integrated management strategy to conserve natural resources for sustainable use, with inter-generational equity concerns. Landscape management demands a variety of responses that are location-specific, in terms of land use activities linked with natural resource management such as the hydrology regime, sustainable soil fertility, biodiversity, and biomass production. These drivers of rural rehabilitation were emphasized whilst dealing with sustainable rural development in the Asian tropics (Ramakrishnan *et al.*, 1994b). What is required is small-scale operations that are location-specific, through community participation, ensured through a variety of institutional arrangements arrived through a participatory mode. In this situation and in all other similar landscape situations, maintenance of the overall sustainability of the systems demands a loosely-coupled management (Table 3), specifically designed to accommodate large variability in ecosystem complexity within a landscape mosaic.

Table 3 *Loosely-coupled landscape management strategy (from Ehrenfeld, 1991)*

- The management plan should stimulate loose coupling within the system, rather than maximize tight coupling. Such a strategy would reduce the system's vulnerability to accidents.
- Avoid manipulative management plans based on large, prioritized, single-protocol, single theory, generalized schemes. Instead, for large protected areas, have only a low-key management plan, if required. Have individualized management for fragmented landscapes, to accommodate singularities and uniqueness of each piece and ensure loose coupling.
- In the ultimate analysis, any management strategy should be flexible, capable of small-scale operations, information-sensitive, and composed of elements that are integrated and yet independent.

REFERENCES

Altieri, M A, Glaser, D L and Schmidt, L L (1990) Diversification of agroecosystems for insect pest Regulation: Experiments with Collards, In: Gliessman, S R (Ed) *Agroecology: Researching the Ecological Basis for Sustainable Agriculture,* Springer-Verlag, New York, pp. 70–82.

Bentley, W R, Khosla, P K and Seckler, K (Eds) (1993) *Agroforestry in South Asia: Problems and Applied Research Perspectives,* Oxford & IBH Publ., New Delhi, p. 367.

Brookfield, H and Padoch, C (1994) Appreciating biodiversity: A look at the dynamism and diversity of indigenous farming practices, *Environment,* **36**: 6–11, 37–45.

Ehrenfeld, D (1991) The management of biodiversity: a conservation paradox. In: Bormann, F H and Kellert, S R (Eds) *Ecology, Economics, Ethics: The Broken Circle,* Yale University Press, New Haven, pp. 26–39.

Faminow, M D (1999) Local knowledge and people's action to reduce deforestation: Slash and agriculture in Nagaland, *Proc. Internat. Workshop: Technology Change in Agriculture and Deforestation,* CIFOR, CATIE, Turrialba, Costa Rica. pp. 24 (mimeographed).

Gliessman, S R (1988) Ecology and management of weeds in traditional agroecosystems, In: Altieri, M A and Liebman, M (Eds) *Weed Management in Agroecosystems: Ecological Approaches,* CRC Press Inc., Boca Raton, Florida, pp. 237–244.

Gliessman, S R (Ed) (1990) *Agroecology: Researching the Ecological Basis of Sustainable Agriculture,* Ecol. Studies **78**, Springer-Verlag, New York, p. 380.

Hughes, J D and Chandran (1998) Sacred groves around the earth: An overview. In: Ramakrishnan, P S, Saxena, K G and Chandrashekara (Eds) *Conserving the Sacred: For Biodiversity Management,* UNESCO and Oxford & IBH, New Delhi, pp. 69–86.

Khiewtam, R and Ramakrishnan, P S (1993) Litter and fine root dynamics of relict sacred grove forest of Cherrapunji in Northeastern India, *Forest Ecol. Manage,* **60**: 327–344.

Letourneau, D K (1990) Two examples of natural enemy augmentation: A consequence of crop diversification. In: Gliessman, S R (Ed). *Agroecology: Researching the Ecological Basis for Sustainable Agriculture,* Springer-Verlag, New York, pp. 11–29.

Litsinger, J A and Moody, K (1976) Integrated pest management in multiple cropping systems. In: Stelly, M (Ed) *Multiple Cropping,* Amer. Soc. Agron., Madson, Wisconsin, pp. 293–316.

Maikhuri, R K and Ramakrishnan, P S (1990) Comparative analysis of the village ecosystem function of different tribes living in the same area in Arunachal Pradesh in Northeastern India, *Agric. Systems,* **35**: 377–399.

Odum, E P (1971) *Fundamentals of Ecology,* W B Saunders Co, Philadelphia, p. 574.

Pimental, D A, Stachow, U, Takacs, D A Brubaker, H W, Dumas, A R, Meaney, J J, Oneil, J A S, Onsi, D E & Corzilius, D B (1992) Conserving biodiversity in agricultural and forestry systems, *Bioscience,* **42**: 354–364.

Pritchard Jr, L, Colding, J, Berkes, F, Svedin, U and Folke, C (1998) *The Problem of Fit between Ecosystems and Institutions.* IHDP Working Paper 2, Internat. Human Dimensions Programme on Global Environmental Change, Bonn, Germany, p. 50.

Ram S C & Ramakrishanan, P S (1992) Fire and nutrient cycling in seral grasslands of Cherrpunji, *Int. J. Wildland Fire,* **2**: 131–138.

Ramakrishnan, P S (1986) Morphometric analysis of growth and architecture of tropical trees. In: *Naturalia Monspeliensia: Colloque International Arbre,* Montpelier, pp. 209–222.

Ramakrishnan, P S (Ed) (1991) *Ecology of Biological Invasion in the Tropics,* National Institute of Ecology & International Science Publ, New Delhi, p. 195.

Ramakrishnan, P S (1992a) *Shifting Agriculture and Sustainable Development of Northeastern India,* UNESCO-MAB Series, Paris, Parthnon Publ, Carnforth, Lancs, UK, pp. 424 (reprinted by Oxford University Press, New Delhi, 1993).

Ramakrishnan, P S (1992b) Tropical Forests. Exploitation, Conservation and Management, *Impact of Science on Society,* **42**(166): 149–162.

Ramakrishnan, P S (1993) Evaluating sustainable development with peoples' participation. In: Moser, F (Ed) Sustainability—Where Do We Stand? *Proc. Intl. Symposium,* Technische Universitat, Graz, Sustria, pp. 165–182.

Ramakrishnan, P S (1994) The jhum agroecosystem in Northeastern India: A case study of the biological management of soils in a shifting agricultural system. In: Woomer, P L and Swift, M J (Eds) *The Biological Management of Tropical Soil Fertility,* TSBF, Nairobi & Wiley-Sayce Publ, Chichester & Exeter, U K.

Ramakrishnan, P S (1996) Ecological functions of biodiversity: The human dimension. In: di Castri, F and Younes, T (Eds) *Biodiversity, Science and Development: Towards a new partnership,* CAB International Wallingford and IUBS, Paris, pp. 114–129.

Ramakrishnan, P S (1995) Currencies for measuring sustainability: Case studies from Asian highlands. In: Munasinghe, M and Shearer, W (Eds) *Defining and Measuring Sustainability: The Biogeophysical Foundations.* U.N. University, Tokyo and World Bank, Washington DC, pp. 193–205.

Ramakrishnan, P S (1996) Conserving the sacred: From species to landscapes, *Nature & Resources,* UNESCO **32**: 11–19.

Ramakrishnan, P S (1998) Ecology, economics and ethics: Some key issues relevant to natural resource management in developing countries, *International Journal of Social Economics,* **25**: 207–225.

Ramakrishnan, P S (1999a) Lessons from the earth summit: Protecting and managing biodiversity in the tropics. In: Dore, M H I and Mount, T M (Eds) *Global Environmental Economics: Equity and Limits to Markets,* Blackwell Publ, Oxford, pp. 240–264.

Ramakrishnan, P S (1999b) The impact of globalization on agricultural systems of traditional societies. In: A K Dragun and C Tisdell (Eds) *Sustainable Agriculture and Environment: Globalization and the Impact of Trade Liberalisation,* Edward Elgar, Cheltenham, UK, pp. 185–200.

Ramakrishnan, P S and Ram, S C (1988) Vegetation, biomass and productivity of seral grasslands of Cherrapunji in Northeast India. *Vegetation,* **74**: 47–53.

Ramakrishnan, P S, Shukla, R P and Boojh, R (1982) Growth strategy of trees and their application to forest management, *Curr. Sci.,* **51**: 448–455.

Ramakrishnan, P S, Saxena, K G, Swift, M J and Seward, P D (Eds) (1993) *Tropical Soil Biology and Fertility Research: South Asian Context.* Himasvikar Publ. 4, Oriental Enterprises, Dehra Dun, p. 147.

Ramakrishanan, P S, Purohit, A N, Saxena, K G & Rao, K S (Eds) (1994a) *Himalayan Environment and Sustainable Development,* Diamond Jubilee Publ, Indian National Science Academy, New Delhi, p. 84.

Ramakrishnan, P S, Campbell, J, Demierre, L, Gyi, A, Malhotra, K C, Mehndiratta, S, Rai, S N and Sashidharan, E M (Eds) (1994a) *Ecosystem Rehabilitation of the Rural Landscape in South and Central Asia: An Analysis of Issues,* Special Publication, UNESCO (ROSTCA), New Delhi, p. 29.

Ramakrishnan, P S, Purohit, A N, Saxena, K G, Rao, K S and Maikhuri, R K (Eds) (1996a) *Conservation and Management of Biological Resources in Himalaya,* G B Pant, Inst of Himalayan Environment & Development, Kosi, Almora, Oxford & IBH Publ, New Delhi, p. 601.

Ramakrishnan, P S, Das, A K and Saxena, K G (Eds) (1996b) *Conserving Biodiversity for Sustainable Development,* Indian National Science Academy, NewDelhi, p. 246.

Ramakrishnan, P S, Saxena, K G and Chandrashekara, U (1998) *Conserving the Sacred: For Biodiversity Management,* UNESCO, Oxford & IBH, New Delhi and Science Publ., Enfield, NH, USA, p. 480.

Ramakrishnan, P S, Chandrashekara, U M, Elouard, C, Guilmoto, C Z, Maikhuri, R K, Rao, K S, Saxena, K G and Shankar, S (Eds) (1999) *Biodiversity, Land Use Dynamics and Traditional Ecological Knowledge,* UNESCO and Oxford & IBH Publ, New Delhi (in press).

Swift, M J and Ingram, J S I (1996) *Effect of Global Change on Multi-species Agroecosystems: Implementation Plan,* GCTE Activity 3.4, Global Change & Terrestrial Ecosystems Report No. 13, GCTE Focus 3 Office, Wallingford, UK.

Swift, M J, Vandermeer, J, Ramakrishnan, P S, Anderson, J M, Ong, C K and Hawkins, B (1994) Biodiversity and agroecosystem function. In: Mooney, H A, Cushman, J H, Medina, E, Sala, O E and Schulze, E-D (Eds) *Functional Roles of Biodiversity: A Global Perspective,* SCOPE Series, John Wiley, Chichester, UK, pp. 261–298.

The Spatio-temporal aspects of Monsoon Floods in India: Implications for Flood Hazard Management

Vishwas S Kale*

ABSTRACT

Many parts of the Indian region are subjected to some of the world's most intense rainfalls and floods during the monsoon season. In certain situations, topographic and meteorological factors locally enhance the magnitude of floods. Consequently, some rivers exhibit a distinctive hydrology, characterized by high flood variability and sometimes large perturbations in the flood magnitude. This poses immense hazards for some densely populated regions of India. For effective flood management and control, therefore, a proper scientific understanding of the causes and the spatio-temporal characteristics of floods is essential.

In the Indian context, the widespread impression is that the flood situation in the country is deteriorating. Although this may be true in terms of flood damages, it does not appear to be true in terms of flood frequency-magnitude and the area affected by floods. An analysis of the flood series (70–100<years), available for some rivers, indicates fluctuations in the flood stage/magnitude on different time scales. Whereas for some rivers these data exhibit a noteworthy increasing or decreasing trend, for other rivers episodic fluctuations in flood magnitude and frequency are indicated.

Since the '50s extensive efforts have been made to minimize flood damages through structural and non-structural measures. However, these

* Department of Geography, University of Pune, Pune 411 007, India

measures have not been able to provide the expected relief in the most flood-prone areas of the country. Form the perspective of flood hydrology and geomorphology, there are three main reasons for this situation:

1. the paucity of long-term hydrological records to evaluate the characteristics of floods,
2. a lack of understanding of the complex physical processes involved in flooding, due to enormous variability in the form and behavioral characteristics of the rivers, and
3. the dearth of data on the engineering, geomorphic, and environmental performance of the flood-protection works.

All available records suggest that the flood phenomena are an inevitable and integral part of the monsoon climate. Therefore, it is felt that river-friendly flood control measures that recognize the environmental value and the geomorphic importance of floods, deserve greater attention in all the future flood control programs in India.

Keywords: India, flood magnitude-frequency, flood hazard, flood variability, monsoon rainfall, trends, palaeofloods, flood management

Introduction

The recent floods in Bihar, Assam and West Bengal, which caused extensive losses in terms of human lives and property, have once again underscored the need for improvement of the scientific understanding of this most recurring, widespread, and disastrous natural hazard. Due to the monsoonal climate with periodic high-magnitude rainfall, and the presence of mighty mountain ranges, some parts of the Indian region are subjected to some of the world's most intense rainfalls and floods during the monsoon season. In order to get a better insight into the enormity and extent of the problem of flood hazard in India, it is necessary to have a comprehensive understanding of the spatio-temporal aspects of the hazard.

In recent years, numerous efforts have been made in this direction. These include the preparation of flood hazard maps, and collection of data regarding the areas prone to floods, the extent of damages and the population affected by floods in different parts of the country. In this regard, it is important to mention the *Flood Atlas of India* prepared by Central Water Commission (CWC), the *Natural Hazard Map of India* prepared by the National Atlas and Thematic Organization (NATMO), and the map of the flood-prone areas given in the *Vulnerability Atlas of India* prepared by the Building Materials and Technology

Promotion Council (BMTPC). These and other maps show that floods comprise a significant hazard in the following area (Central Water Commission Publication, 1996; Dhar *et al.*, 1986; Kale, 1997; Kale, 1998) (Figure 1):

1. Sub-Himalayan region and the Ganga plains
2. Brahmaputra Valley
3. Punjab Plains
4. Mahanadi-Godavari-Krishna-Kaveri Delta plains
5. Lower Naramada-Tapi-Mahi Valleys

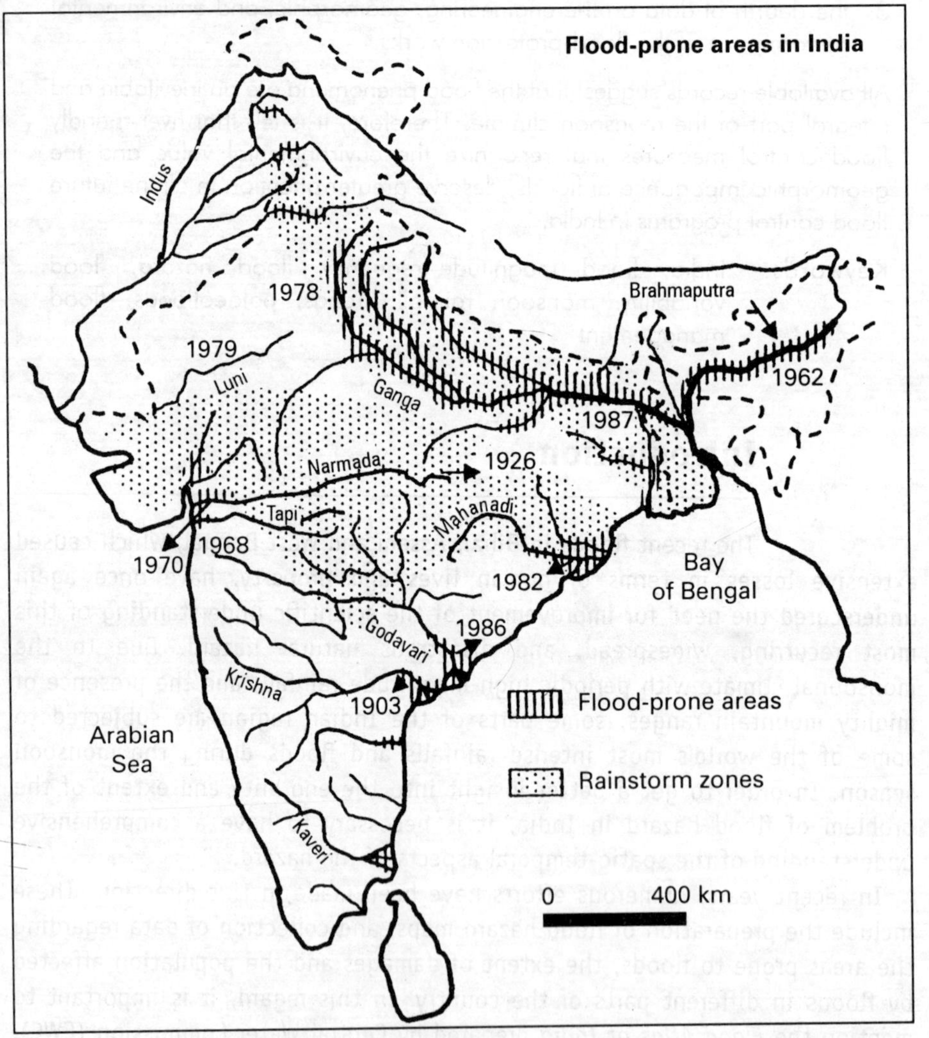

Figure 1 *Major flood-prone areas (Central Water Commission Publication, 1996) and zones of rainstorms in India (Nandargi, 1996). The years of occurrence of highest floods on different rivers are also given.*

The first three areas, drained by the Himalayan rivers and their tributaries, represent the most acutely flood-prone regions of the country. In terms of flood behavior, the Himalayan rivers are very unusual when compared with mountainous rivers elsewhere. Potent monsoonal rains, very high basin relief, steep and unstable hillslopes, and tectonic instability are some of the critical geo-environmental factors that have been responsible for the unique flood regime characteristics of the rivers.

Since 1953, data regarding flood damages in different states and for the entire country have been collected (Central Water Commission Publication, 1996). According to the estimates of the National Commission of Floods or *Rashtriya Barh Ayog* (RBA), as much as 40 million hectares (mha) of land is prone to floods in the country (Central Water Commission Publication, 1996). Most of the flood-prone areas, identified by RBA and other agencies, lie in the Ganga (Uttar Pradesh, Bihar and West Bengal) and the Brahmaputra (Assam) Basins. In these basins about 15.36 and 3.82 mha area is prone to floods, respectively (Central Water Commission Publication, 1996). The other frequently affected states are: Haryana, Himachal Pradesh, Punjab, Orissa, Andhra Pradesh, and Gujarat.

Even though the maximum flood-prone area in the country is 40 mha, floods annually affect an average area of about 7.6 mha. This area has varied significantly in the last five decades, from as low as 1.46 mha in 1965 to as high as 17.5 mha in 1978 (Central Water Commission Publication, 1996). Floods annually affect approximately 32 million people. This number has varied from 3.61 million in 1965 to 70 million in 1978 (Central Water Commission Publication, 1996).

Although floods occur with an unfailing regularity in some states, such as Uttar Pradesh, Bihar, West Bengal and Assam, the geographical area affected by floods does not remain constant every year, but varies considerably from year to year. This temporal variability in the flood-affected area since the early 1950s is depicted in Figure 2. The graphs exhibit a significant increase in the flood-affected area between 1970 and 1980 in the Ganga Basin. In comparison, the flood-affected area in the Brahmaputra Basin has remained close to the average for the last four-and-a-half decades with three short periods of large floods, viz., 1954–55, 1972–74, and 1987–88 (Kale, 1998). These plots therefore, do not provide any evidence of an increase in either the flood-prone area or the frequency of large floods over the last few decades (Kale, 1998; Hofer, 1998). This is contrary to the widespread belief that the flood situation in the country is worsening (Hofer, 1998).

Another noteworthy characteristic of the flood pattern reflected by the graphs (Figure 2) is the asynchronous nature of large floods in the four most flood-prone states of the country. This is to say that, at least in the last five

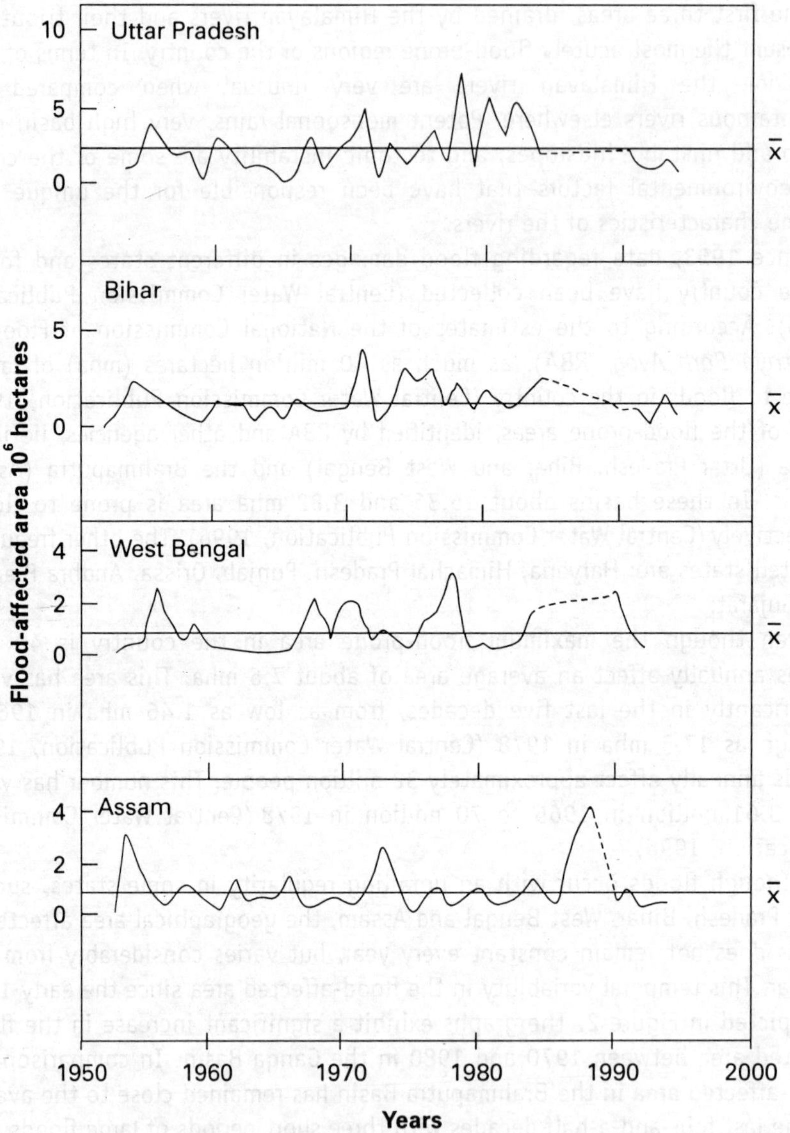

Figure 2 *Year-wise distribution of the area affected by floods in the Ganga and Brahmaputra Basins (Central Water Commission Publication, 1996, CSE, 1991; Chaphekar et al., 1985). Dashed line represents missing data. $\bar{x}$ = average flood affected area.*

decades, there is no evidence to suggest that there is a tendency for large floods to occur at around the same time in different parts of the Ganga Basin and/or Brahmaputra Basin. The non-synchronicity in the occurrence of floods in the Ganga-Brahmaputra plains suggests that not only the flood phenomena are regionally limited (Hofer, 1998), but also the flood-generating hydro-meteorological conditions vary spatially (Kale, 1998). This has important implications for the prediction and management of the flood hazards in the most flood-prone regions of the country.

Flood Hydrology of Indian Rivers

Despite the problems of hydrological data, the distinctive hydrological features of Indian rivers are now reasonably well established (Kale, 1998; Hofer, 1998; Sakthivadivel *et al.*, 1978). Numerous hydro-geomorphic studies in the last four decades indicate that the rivers are characterized by:

a) high flood magnitudes
b) high flood variability
c) significant long-term variations in the flood regime conditions

Every year about six to eight major floods occur in one or the other part of the country (Dhar, 1986). Although the rivers have two to three months of high flows and floods can occur during any part of the monsoon season, the annual flood peaks are generally recorded in the months of August and September. The annual peak floods on large rivers range between 6,000 and 80,000 m^3/s. These are approximately 1.3 to 4 times greater than the mean annual flood discharges (Table 1) and rank among the global highest for their respective drainage areas.

Whereas single-peak flood events during the monsoon season are a common feature in case of most of the rivers, sometimes successive flood-generating meteorological conditions may give rise to multiple peak events. The 1970 floods on the Narmada River, for example, were associated with such a situation. Multiple floods are also relatively frequent in the Ganga plains. Single-peak events are commonly observed on smaller rivers or on rivers flowing mostly through arid and semi-arid areas.

Short-lived, but highly damaging, flash-floods are generally produced by intense convectional storms during the post- and pre-monsoon seasons, and sometimes during the monsoon season. Drainage basin characteristics (steep slopes, high drainage density) also play an important role in the generation of flash-floods.

There is considerable inter-regional variability in both the annual monsoon flows and the peak flood discharges (Table 1). This is primarily due to the marked variation in the regional pattern of monsoon rainfall, with some basins receiving higher rainfall in the catchment areas than the others. As a result of this there is great diversity in hydrological characteristics within the monsoonal region, with some rivers revealing ephemeral characteristics (such as

Table 1 *Flood discharge characteristics of some Indian rivers (Sakthivadivel et al., 1978; Goswami, 1998; Goswami, 1985; Seth et al., 1982)*

Name of the river and the gauging site	Mean flood discharge m^3/s	Highest flood discharge on record m^3/s	Highest to mean flood discharge ratio
Ravi at Madhopur	4351	17471	4.02
Tapi at Kathore	10748	41700	3.88
Pennar at Nellore	3623	13394	3.69
Yamuna at Tajewala	4400	15942	3.62
Ganga at Raiwala	6055	19133	3.16
Godavari at Dowleshwaram	29207	79990	2.74
Baitarni at Akhupada	3687	9203	2.49
Narmada at Garudeshwar	28133	69400	2.47
Kaveri at K R Sagar	2641	6205	2.35
Sutlej at Bhakra	3941	9203	2.33
Damodar at Rhondia	6328	14761	2.33
Krishna at Vijayawada	14775	33600	2.27
Kosi at Sunakhambikhola	8178	17445	2.13
Sone at Dehri	17873	34235	1.92
Brahmaputra at Pandu	48161	72794	1.51
Mahanadi at Naraj	29269	42334	1.44
Ganga at Farakka	55776	72915	1.31

the Luni River) and other revealing more sustained high flows during the monsoon season.

One widely used measure of the flood variability is the flash flood magnitude index (FFMI). The index is the standard deviation of logarithms to the base of 10 of the annual maximum series (Baker, 1977). The FFMI values for some river gauging stations are given in Table 2. Rivers with higher flash flood magnitude indices such as the Damodar, Teesta, Upper Yamuna, and Luni reveal large variability in flood discharges, as that of other rivers known for large floods such as the Lower Ganga and Brahmaputra. The table, however, reveals that the FFMI values for most large Indian rivers are lower than those in other parts of the world (McMahon *et al.*, 1992; Erskine *et al.*, 1988). This implies that, by and large, the inter-annual variations in the flood magnitudes are relatively lower than other world rivers. This in turn suggests that, as compared to other world rivers, large magnitudes are recorded relatively frequently on the Indian rivers.

Over the last fifty years many Indian rivers have encountered some of the largest floods in this century. These include the 1962 flood on the Brahmaputra,

Table 2 *Flash Flood Magnitude index for selected rivers in India (Kale, 1998; Sakthivadivel, 1978; Goswami, 1998; Goswami, 1985; McMahon, 1992)*

River	Length of record	FFMI
Damodar at Rhondia	36	0.31
Yamuna at Tajewala	55	0.21
Teesta at Teesta bridge	20	0.21
Narmada at Garudeshwar	42	0.20
Krishna at Vijaywada	96	0.20
Tapi at Bhuranpur	25	0.19
Godavari at Dowleshwaram	68	0.18
Ganga at Raiwala	97	0.15
Sutlej at Bhakra	115	0.09
Brahmaputra at Pandu	43	0.08
Ganga at Farakka	22	0.08
Luni at Gandav	24	1.3*
World Rivers	–	0.28

* = After deleting years with zero discharge

the 1978 and 1987 floods on the Ganga, the 1979 flood on the Luni, the 1968 flood on the Tapi, the 1970 flood on the Narmada, the 1982 flood on the Mahanadi, and the 1986 flood on the Godavari (Central Water Commission Publication, 1996; Kale, 1998). These unusual floods represent large perturbations of the magnitude–frequency distribution of flood flows.

Types of Floods in India

Floods of different types occur in India. These include the rainfall, rainstorm, coastal and dam-failure floods. Snowmelt floods are not very important in the Indian context.

a) ***Rainfall floods*** are the most frequent in the annual flood series for most rivers of the subcontinent. Rainfall floods result from heavy or intense precipitation in association with active to vigorous monsoon conditions for a number of days (Dhar *et al.*, 1998) (for example, the extraordinary flood of 1962 on the Brahmaputra River).
b) ***Rainstorm floods*** are generated by excessively heavy rainfall associated with lows, depressions and cyclonic storms originating over the Bay of Bengal or

the adjoining coastal belt (Dhar *et al.*, 1998). Cyclonic storms contribute significantly to the total precipitation and produce devastating floods. Some of the worst floods recorded in recent times, such as the 1968 flood on the Tapi and Teesta, the 1970 flood on the Narmada, etc., were associated with cyclonic storms that formed over the Bay of Bengal or the adjoining coastal belt.

c) ***Coastal floods*** connected with storm surges occur in some parts of the eastern coast of India, and are produced by severe cyclonic storms originating over the Bay of Bengal. Some of the most severe storm surges were recorded in 1970, 1977 and 1982 in Orissa and/or Andhra Pradesh.

d) ***Dam-failure floods*** are the most disastrous floods, and are usually the result of the failure or breaching of natural or man-made dams (Table 3 and 4). In this century there have been more than a dozen anomalous floods that were related to failure of man-made dams (Table 4).

Table 3 *Some well-known major floods caused by the failure of natural dams (Kale, 1997; Khan, 1969; CSE, 1991)*

River	Year of dam-failure flood
Indus	1841 and 1858
Shyok	1926 and 1929
Birehganga	1893
Alaknanda	1970
Bhagirathi	1978
Sutlej and Bagmati	1993
Malipa Gad and Kali	1998

Table 4 *Some major floods caused by the failure of man-made dams (Kale, 1997; CSE, 1991)*

Name of the Dam/State	Year of dam-failure flood
Tigra (Madhya Pradesh)	1917
Kundali (Maharashtra)	1925
Kadam (Andhra Pradesh)	1958
Panshet (Maharashtra)	1961
Kharagpur (Bihar)	1961
Morvi (Gujarat)	1979
Jaswantnagar (Rajasthan)	1979
Chandora (Madhya Pradesh)	1991

Large floods have also originated from the breaching of natural dams formed by landslides, moraines and glacial ice in different parts of the Himalayas (Table 3). For example, in August 1998, flash floods occurred at Malpa due to the breaching of a natural dam created across Malipa Gad and Kali River. The debris dam was created because of a massive landslide and rockfall, following incessant rains in the area.

Of all the four types of floods mentioned above, the dam-failure floods are most devasting, because they are least predictable.

Causes of Floods

Meteorological Causes

There is substantial inter-annual variability in the monsoon rainfall over the Indian region. With high variability in monsoon rainfall, there are inevitably droughts or floods. Therefore, it is not surprising that India is often referred to as the land of droughts and floods. It has now been well established that this inter-annual variability in the monsoon rainfall, is teleconnected to large scale phenomena such as the El Niño/Southern Oscillations (ENSO) (Bhalme *et al.*, 1984; Parthasarathy *et al.*, 1991; Whetton *et al.*, 1994).

Even during a normal monsoon year, some parts of the country receive more rainfall than the others. Therefore, for a given year, the occurrence and distribution of large floods are determined by the amount and distribution of heavy to excessive rainfall. Exceptionally heavy to heavy rainfall during the monsoon season is generally associated with (Dhar *et al.*, 1998; Ramaswamy, 1987; Rakhecha *et al.*, 1996):

a) Monsoon depressions and cyclonic storms originating over the Bay of Bengal and the adjoining coastal belt, and the Arabian Sea
b) Orographic lifting along mountain barriers
c) Breaks in the monsoon

Large floods on most Indian rivers are a direct result of intense cyclonic storms and depressions. In general, exceptionally heavy rains are associated with slow moving and stationary low-pressure systems (Dhar *et al.*, 1998; Nandargi, 1996). The flood-generating rainstorms are mainly confined to two major zones (Nandargi, 1996). The first zone is located over the Ganga Basin and the Punjab plains, and the second zone includes the central India as well as the northern half of the Peninsular India (Figure 1).

Geomorphic Causes

Landslides (initiated by intense rains and earthquakes) or surging glaciers block rivers, and cause massive floods when they fail. Numerous examples of breach floods have been recorded on many Himalayan rivers such as the Indus, Shyok, Lihut, Sutlej, Bhagirathi, Alaknanda, Birehganga, Teesta, Subansiri, Malipa gad, Kali and many others.

Avulsion, channel migration, meander growth, and natural changes in the elevation of river channels (due to excessive siltation) are some of the other geomorphic causes (Kale, 1998) of large floods in parts of the Ganga and Brahmaputra Basins. Sometimes, major changes in the channel morphology are brought about by earthquakes. Such changes were reported after the 1897 and 1950 earthquakes in Assam (Goswami, 1985).

Anthropogenic Causes

In recent decades, many parts of the Ganga–Brahmaputra region have experienced devastating floods. There is a tendency among some scientists, environmentalists and even the local people to attribute this to changes in land use, large-scale deforestation in the catchment areas and building activity in the flood plains (CSE, 1991; Chaphekar *et al.*, 1985). Despite such vociferous claims, recent investigations have not been able to scientifically prove the connection between environmental degradation in the Himalayas and increased flood damages in the Ganga and Brahmaputra basins (Hofer, 1998; Messerli *et al.*, 1995; Brammer, 1990).

However, there is little doubt that the failure of man-made dams and other structures across rivers have contributed to aggravating the flood problems in some areas, and have been responsible for some of the most devastating floods in this century. Some of the best-known examples include the dam-failure floods of Panshet in 1961, Morvi in 1979 and Chandora in 1991 (Table 4). In addition to these, severe flood damages have also occurred due to the breaching of embankments and other flood-protection structures.

Decade Scale shifts in Flood Regime Conditions

It is increasingly being recognized that long-term changes in rainfall regime at the regional scale can significantly affect the magnitude and frequency of floods. Several studies based on hydrological data from rivers from around the world have now provided evidence of rainfall-related changes in the flood activity, and rainfall-driven alternating periods of low and high flood activity (Erskine *et al.*, 1988; Probst *et al.*, 1987; Burn *et al.*, 1993; Kale, 1999).

The probability of detecting shifts or changes in the stream-flow records on the decadal or century scale is greater from longer records. Although such long records of rainfall are available for many stations in India (for some stations since 1813), there are only a few river gauging sites where some records are available from the middle or the end of the last century (Kale, 1998).

The inherent inter-annual epochal variability in the monsoon rainfall with the periods 1877–1898 and 1932–1964 being above-average (humid) and the periods 1900–1932 and 1965–1987 being below-average (dry) have been now well established (Kripalani *et al.*, 1997; Kale, 1999; Wright, 1989; Gregory, 1989). An analysis of the gauge records available for some rivers demonstrates quite a good correspondence between the periods of increased frequency of floods (low floods) and increased average-rainfall (below-average rainfall) at the macro-regional/all-India level (Kale, 1998; Kale, 1999). This correspondence suggests that long-term changes in the rainfall regime over the Indian region have strongly influenced the spatio-temporal distribution of large floods.

In the Deccan Peninsula, low flood activity has been documented between 1901 and 1940 and flood-dominated conditions have been documented between the 1940s and 1960s (Kale, 1999) (Figure 3). This trend is broadly consistent with the long-term variability in the all-India monsoon rainfall and SST index of ENSO (Wright, 1989). Parthasarathy *et al.* and Krishnamurthy and Goswami have shown that the period during 1901–1940 was dominated by stable "zonal monsoon", and the period during 1871–1900 and after 1940 were characterized by less stable "meridional monsoon". Therefore, the clustering of large flood events between 1940s and 1960s as well as during the last few decades of the 19th century appears to be strongly linked to secular changes in the monsoon rainfall.

Due to the paucity of long-term gauge records for most flood-prone Himalayan rivers, such as the Lower Ganga, Brahmaputra, Gandak, Kosi, Teesta, Manas, Subansiri, etc., it has not been possible to identify rainfall-related periods of low and high flood activity. Therefore, any method of establishing the periodicity of floods for such flood-prone rivers would have considerable practical implications in terms of preparations for and responses to future flood disasters.

An estimate of the future flood conditions is required for either flood prediction (for design of dams, bridges river channel modifications etc.) or for flood forecasting and warning. To accurately predict the occurrence of large floods, engineers and hydrologists have traditionally relied on flood-frequency analysis (Baker, 1994; Baker *et al.*, 1995). However, in order to obtain valid and

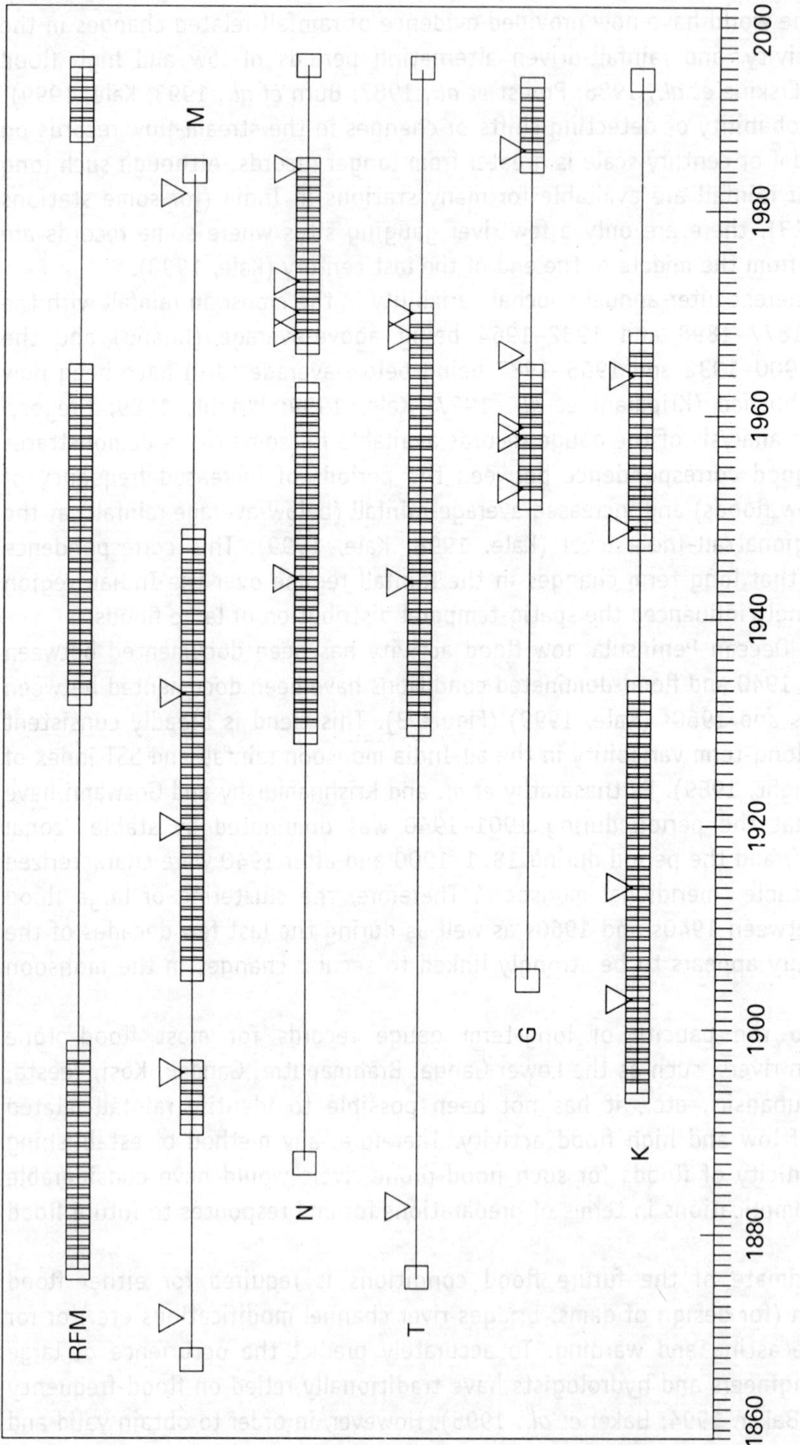

Figure 3 *Periods of the high (above-average) and low (below-average) floods in the Deccan peninsula (after Kale 1999).*

reliable estimates from flood-frequency analysis, it is necessary to first demonstrate that the hydrologic characteristics of the rivers have not changed during the period of record (i.e., stationarity of flood series). Since there is enough evidence now to suggest that most rivers are characterized by rainfall-driven periods of low and high floods of several decades, the flood frequency analysis is subject to large error when estimating the future flood conditions. Therefore, sequences of wet years or dry years must be taken into account when attempting to predict flood peaks for either flood prediction or flood forecasting.

Long-term Flood Variability

In the fifties, sixties and seventies several extreme flood events occurred, with 1953, 1959, 1961, 1968, 1970, 1973 and 1978 being particularly notable. This prompted some scientists, environmentalists and even local people to consider the recent increase in floods and flood damages as the result of increased anthropogenic activities. However, experience from scientific studies of the hydrological data available for the last 100–150 years indicates that this is not entirely true.

In order to verify the widespread belief that the frequency of abnormal floods has increased substantially in the last few decades, Kale attempted to evaluate the long-term trends or changes in the hydrological records as well as in the flood-prone area by using the test of Mann–Kendall (Hollander *et al.*, 1973). This non-parametric method has been used by many earlier workers to quantify the direction and magnitude of trends in the streamflow and rainfall records (Chiew *et al.*, 1993; Marengo, 1995). His analysis indicates significant decreasing trends for the Upper Ganga, Krishna, Brahmaputra and Sutlej Rivers (Table 5). The test, however, does not reveal any statistically significant increase or decrease in the discharge stage of the Yamuna and Mahanadi Rivers, the all-India monsoon rainfall and the area affected by floods (Table 5). Similarly, no significant change in the flood-affected area in the Upper, Middle and Lower Ganga Basin (UP, Bihar and West Bengal, respectively) as well as in the Brahmaputra Valley (Assam) was indicated by the test. These results, therefore, provide a weak support to the widespread view that floods in India (especially in the Ganga and Brahmaputra Basins) are progressively increasing.

An equally important aspect in flood hazard research is the knowledge of the changes in the pattern of flooding in the near future. With the existing level of understanding of the flood phenomenon it is nearly impossible to predict the direction and magnitude of change. It is, however, possible to estimate the percentage change required in the future data series before it can be considered

Table 5 *Nature of changes/trends in hydrological records based on Mann–Kendall test (Kale, 1998)*

River	Site	Parameter	Time-span	Tau	Trend or change
Godavari	Dowleshwaram/ Polavaram	Peak discharge*	1905–1992	+0.1771	I,SS
Narmada	Baroach	Peak stage	1887–1994	+0.1822	I, SS
Sutlej	Bhakra	Peak discharge	1912–1995	+0.1987	I, SS
All-India	Area affected by floods	Area in mha	1953–1994	+0.9536	I, NS
All-India	Population affected by floods	Population in millions	1953–1994	+3.7823	I, SS
Mahanadi	Naraj	Peak Stage*	1968–1984	–0.0057	D, NS
Yamuna	Tajewala	Peak discharge	1910–1985	–0.0081	D, NS
All-India	Monsoon rainfall	Rainfall	1871–1990	–0.0118	D, NS
Krishna	Vijayawada	Peak discharge*	1894–1993	–0.1851	D, SS
Ganga	Raiwala	Peak discharge	1881–1977	–0.2779	D, SS
Brahmaputra	Pandu	Peak discharge	1955–1997	–3.3384	D, SS

D = decreasing trend; I = increasing trend; SS = statistically significant at 0.05 level; NS = statistically insignificant at 0.05 level; * = data missing for some years.

to be indicative of a major change in the hydro-climatic conditions. Several workers have used the students *t*-test to determine the percentage change required in the mean of the future flood series before it can be considered to be significantly different from the past hydrological records (Chiew *et al.*, 1993; Marengo, 1995).

The application of the test reveals that in the Indian situation about 6% change in the all-India monsoon rainfall, about 25 to 30% change in the flood-affected area and flood peaks is required before the mean of the next ten years can be considered as statistically different (Kale, 1998) (Table 6). The table further indicates that major shifts in the rainfall and flood regime conditions are required in the next two or five decades, before the shifts can be considered to be statistically significant. Therefore, monitoring the changes in the hydrological regime conditions is of critical importance for understanding the various natural and anthropogenic forces on the hydrological system, as well as for improving the predictability of the flood pattern in the near future.

Table 6 *Percentage change required to identify statistically significant change in mean discharge/stage/rainfall/flood-affected area (Kale, 1998)*

River/area	Site/parameter	Mean	Coefficient of variation (%)	Percentage change required in the mean at 95% of confidence level Years		
				10	20	50
Ganga	Raiwala – Q	6732.64	37.20	25	18	13
Yamuna	Tajewala – Q	5915.96	58.77	39	30	21
Mahanadi	Narja – stage*	26.57	2.87	2	1	1
Narmada	Baroch – stage	8.41	17.01	11	8	6
Godavari	Dowleshwaram/ Polavaram – Q*	31579.45	46.33	30	23	16
Krishna	Vijaywada – Q*	14015.24	39.27	26	19	14
Brahmaputra	Pandu – Q*	49646.27	18.15	14	12	10
All-India	Monsoon rainfall	851.03	9.92	6	5	3
All-India	Flood-affected area	7.56	3.48	33	25	19

Peak discharge (Q) in m^3/s; peak stage in m; rainfall in mm; area in mha;
* = data missing for some years.

Flood Management Problems

Floods constitute a rather severe natural hazard in India, especially in the Ganga–Brahmaputra Basins. Therefore, several programs to control floods were initiated after the launching of the National Flood Control Program in 1954 and after the constitution of the Rashtriya Barh Ayog (RBS) or National Flood Commission in 1976 (Central Water Commission Publication, 1996; CSE, 1991). The programs primarily included engineering structural measures (Table 7). As a result of the flood control programs implemented in the last few decades, about 14.4 million hectares of land has benefited (Central Water Commission Publication, 1996).

The flood damages have also been reduced to some extent through the use of the flood forecasting system, which has been in operation in some areas for the last 15 years or so (Central Water Commission Publication, 1996). In addition to these non-structural measures, satellite data have also been used very effectively in the last few years for flood mapping and flood-damage assessment.

However, it is important to note that a damaging flood is not just excess rainfall, but also a complex phenomenon requiring deeper understanding. Therefore,

Table 7 *Flood management programs in some flood-prone states (up to March, 1993) (Central Water Commission Publication, 1996)*

State	Maximum flood-prone areas (mha)	Town/ village protection works	Villages raised or protected in km	Length of embankments in in km	Length of drainage channels
Uttar Pradesh	7.3	64	4511	1811	3593
Bihar	4.3	47	-	2788	365
Punjab	4.1	3	-	1370	6622
Himachal Pradesh	3.9	-	-	58	11
Assam	3.8	89	-	4566	957
West Bengal	3.8	48	-	1184	1648
Haryana	2.4	180	90	662	3922
Orissa	1.4	14	29	1068	131
Andhra Pradesh	1.4	52	21	572	13569
Gujarat	1.4	229	30	952	271
India total	36.7	906	4705	16200	32003

in spite of many efforts to control and manage floods, flooding continues to be the single most destructive natural hazard in the country, especially in the Ganga and Brahmaputra Basins. In this context, the primary problems are (Kale, 1997; Kale, 1998; Baker *et al.*, 1995):

a) Lack of long, continuous and reliable hydrological records to analyze and evaluate the flood regime characteristics of the monsoonal rivers
b) The overall difficulty in understanding the complex flood processes due to enormous variability of river types
c) Lack of quantitative data on the flow phenomena and sediment transport during extreme floods with annual exceedence probabilities of less than 10^{-2}
d) Lack of multivariate data on the gemorphic and environmental performance of the flood protection works, to evaluate the consequences of flood control structures.

The inadequacy of hydrological data (both spatial and temporal) is a serious obstacle in understanding flood regime characteristics of the monsoonal rivers, and assessment of flood risk. Since the distribution of large floods in time is highly variable, a few decades of gauging records is not generally considered as a sufficiently representative period for understanding the flood characteristics of large rivers. Further, for a flood to be considered as severe or catastrophic, it must be established that the event is likely to occur with a statistical frequency of once in 50, 100 or more years. This kind of estimation requires at least fifty years of data. Such long, reliable and continuous instrumental records are rarely available in India. In such situations, the use of historical and geological records has often been advocated as a means of extending the short-term instrumental records.

Unlike Egypt (the flood records of the Nile River) or China (*Chinese Atlas of Floods and Droughts between 1470 and 1979*), no continuous historical records of floods or rainfall are available in India although some historical information is available in the *Imperial Gazetteer of India* and other historical records, the flood records are not as extensive, continuous and reliable as the records of droughts and famines in India. Therefore, in recent years an attempt have been made to infer about the past floods (palaeofloods) through the scientific study of slackwater flood deposits (Figure 4) or palaeostage indicators (SWD-PSI). Slackwater sediments offer valuable information about large floods that have occurred before the time of systematic observation or measurement (Baker, 1994). In the last decade, numerous good-quality SWD-PSI records of ancient high-magnitude floods have been identified on some peninsular Indian rivers (Ely *et al.*, 1996, Kale, 1999; Kale, 1997) (Table 8). These records indicate a period of significantly reduced frequency of large floods during the late Medieval and the Little Ice Age (around 1500 AD to late 1800s) and increase in the magnitude and frequency of large floods in the post-1950 period (Kale, 1999) (Figure 5). The analysis and interpretation of the palaeoflood records in the central Narmada Basin have revealed four palaeoflood periods during the last 2000 years. Two sequences of frequent, moderate floods date between 0–400 AD

Table 8 *Summary of palaeoflood records in central and western India (Marengo, 1995)*

River	Site	Max. thickness of SWD m	Number of floods	Length of the record in 10^3 yrs	Number of large post-1950 floods
Narmada	Sakarghat	6.5	26	2.0	3
Narmada	Chahin Nala	5.0	11–15	1.7	4–5
Narmada	Chota Tawa	4.0	14–15	1.4	4
Narmada	Ramghat	2.5	18–19	0.6	4
Godavari	Pochavaram	4.5	32	2.1	1
Krishna	Srisailam	10.5	>37	1.0	1
Tapi	Ghuttigarh-Khapa	3.0	13–16	0.4	1–5
Luni	Bhuka	3.0	17–22	1.0	1

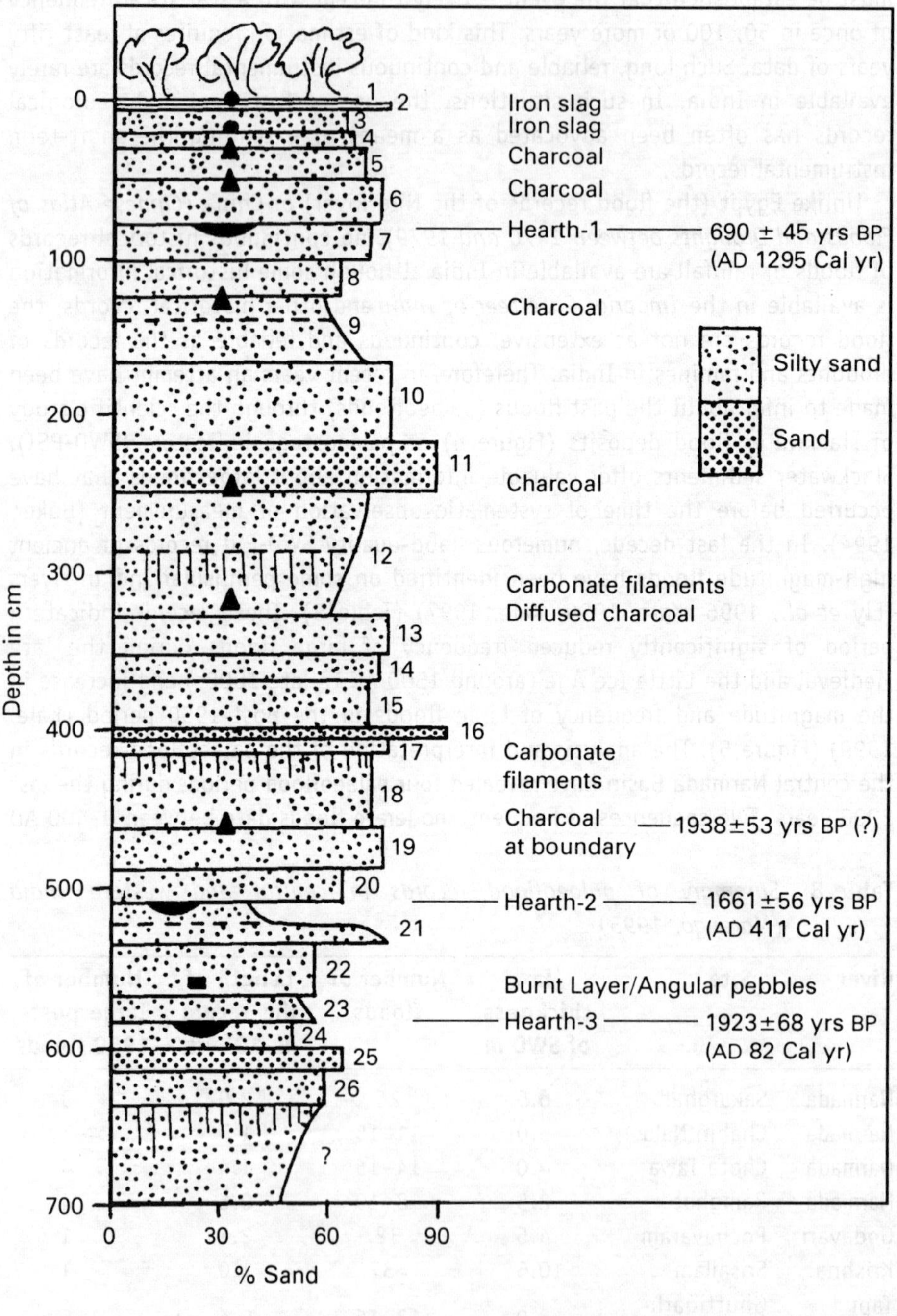

Figure 4 *Palaeoflood deposits exposed on the left bank of the Narmada River at Sakarghat near Punasa (Imperial Gazetter of India, 1908).*

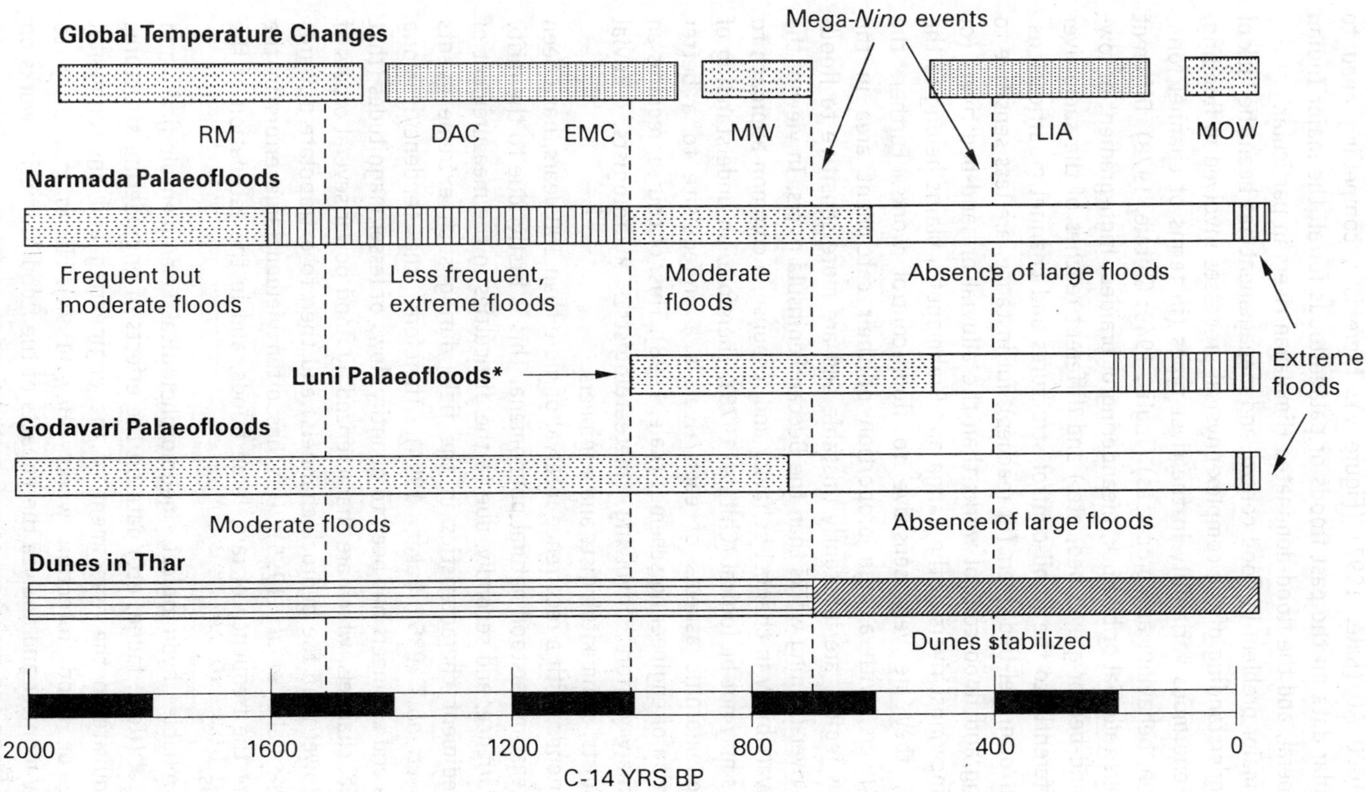

Figure 5 *Periods of high and low floods during the last 2000 years, based on the interpretation of palaeoflood records in central and western India (Marengo, 1995); RM = Roman Warm; DAC = Dark Ages Cold; EMC = Early Medieval Cool; MW = Medieval Warm; LIA = Little Ice Age; MOW = Modern Warm; * = OSL dates.*

and 1000–1400 AD, and a period of less frequent, but more extreme floods date from 400–1000 AD (Kale, 1997) (Figure 5). There is a compelling need to acquire similar data on the past floods or palaeofloods for all the major Indian rivers in general, and the flood-dominated Himalayan rivers in particular.

Another major problem in flood control and management has been the lack of scientific understanding of the complex physical processes involved in flooding, due to the enormous variability in the river types (in terms of channel form, configuration, behaviour and processes) (Inglis, 1949; Chitale, 1978). Different types of rivers (alluvial or bedrock, meandering or braided, high-gradient or low-gradient, sand-bed or gravel-bed, etc.) and different reaches of the same river respond differently to the flood-control structures and channel modifications. Bedrock and/or gravel-bed rivers (or reaches), for instance, are less sensitive to river training and flood-control works than the alluvial or sand-bed rivers (or reaches). Since most rivers of the Ganga and Brahmaputra plains belong to the latter type, they are very sensitive to flood-control works. Further, the Himalayan rivers, with a high proportion of their catchment area in the mountainous region, are inherently unstable and are more sensitive to flood-control and river-training works than the Deccan Peninsula rivers. In view of the substantial variability in channel behavior and response, a common approach to all the rivers may not be logical (Chitale, 1978). Therefore, an understanding of the hydro-geomorphic aspects of each river is a prerequisite to a better implementation of engineering schemes (Kale, 1998). This, of course, requires an interdisciplinary effort involving meteorologists, hydrologists, fluvial gemorphologists, sedimentologists and engineers.

Extreme floods, with a recurrence interval of more than 100 years, have been one of the least understood natural phenomena. This is firstly, due to the rarity of their occurrence, and secondly, due to the difficulties in the measurement of flow and sediment characteristics in the field during such extreme events (Chaphekar *et al.*, 1985; Baker, 1994). Therefore, the tendency among hydrologists and engineers has been to quantify flows of lesser magnitudes (that too in alluvial channels, which are characterized by flood power several orders of magnitude lower than the bedrock channels) and then to extrapolate upwards (Baker, 1994; Baker *et al.*, 1995). This has often yielded large errors in the estimation of the frequency of rare, large floods, and in the estimation of the design floods.

Finally, reliable hydrological, geomorphological, sedimentological and environmental (flora, fauna, etc) data on the effects of engineering structures are a prerequisite to the assessment of the engineering and environmental performance of flood-protection works (Hey, 1994). It is necessary to continuously monitor and analyze the impacts of the flood protection works on the channel forms and processes as well as the riverine environment. This would be helpful in designing more river-friendly flood-protection works on rivers in similar hydro-geomorphic regions. Unfortunately, this is one of the most neglected aspects in the flood management programmes in India. Therefore, it is essential to generate reliable, multi-disciplinary data to scrupulously evaluate the geomorphic and environmental consequences of flood-protection works.

Flood Hazard Management

The International Decade of Natural Disaster Reduction (IDNDR) concluded in July 1999 at Geneva, but there appears to be no end to the problem of floods in the most of the flood-prone areas of the world, including India. Despite of the numerous structural and non-structural measures adopted, the flood hazard has continued, by and large, with the same severity and frequency. Needless to say that there are some deficiencies in the existing disaster management programmes, and hence some rethinking is required. The following are some of the issues that should be considered while formulating the future disaster management program.

The bulk of India's population lives on riverine and coastal plains, which are often prone to floods. Floods in these areas affect large numbers of people and their livelihood. Therefore, in such areas, an integrated disaster management approach is essential that minimizes loss of life, injury and loss or damage to property. Disaster management activities usually include disaster prevention, mitigation, preparedness, response, recovery and rehabilitation. Measures that are multi-pronged and river-friendly are usually needed to provide maximum protection from the flood hazards. Structural measures, including dams, levees, channel alterations and other protective works, are usually designed to provide protection against some specific level of flooding. In terms of potential damages, environmental impact and other factors, the level of protection is an important aspect in riverine and coastal plains. For all types of structural measures in areas vulnerable to flooding, the minimum standard of protection should be last against the 100-year flood. Any standard lower than this has a low level of protection. Often, however, lower levels of protection are favored because of the very high overall initial cost of high-level protection works, with disastrous consequences. A large number of cases of levee failures and dam failures (Table 4) in the last few decades undoubtedly suggest that the associated floods were greater than their design flows. The lack of long, continuous and reliable hydrological records is a major constraint in estimating the minimum standard of protection. It is therefore, essential to improve, modernize, and expand the existing network and rainfall and stream gauging stations, and develop a comprehensive database. These are vital for flood management in the future.

Second, the rivers of the Ganga–Brahmaputra plains are known for frequent and sometimes sudden channel migrations. The existing structural measures have proved to be unsuccessful in such cases. Therefore, there is a need to develop technologies and alternative methods to control the flood hazards on migrating streams and rivers.

Third, the non-structural measures, such as flood forecasting and warning system, flood risk analysis and mapping, as well as disaster relief and preparedness plans, are likely to be significantly more effective in reducing flood damages as well as are likely to be more cost-effective than the structural measures, especially in case of the rivers of the Ganga–Brahmaputra Plains. Accurate and timely information are the keys to effective disaster management. In recent years, remote sensing techniques (particularly the satellite-based radar imaging systems) have demonstrated their usefulness in monitoring and mapping major floods as well as for assessing flood damages in the most flood-prone area (Sharma *et al.*, 1996; Rodriguez *et al.*, 1993). The essential characteristics of the remote sensing data are: all-weather coverage, high spatial resolution (< 10 m), high frequency of observations (at least once a week), and a short data delivery time (one day or less). The existing satellite systems have the greatest potential for disaster mitigation and preparedness activities. The use of satellite data and Geographical Information System (GIS) can facilitate in real time data management, generation of flood maps via GIS, and daily forecasting of the flood situation for 24, 48 and 72 hours.

Four, in connection with the disaster management it is important to recognize that a damaging flood is not just the excess of rainfall, but also a complex phenomenon requiring a deeper understanding of the process of flooding. Whereas the design of measure to control floods is the role of engineering, the understanding of floods is the role of science (Baker, 1994; Baker 1998). The existing scientific knowledge of flood phenomena can be advanced by generating reliable data on the meteorological, hydrological and geomorphological aspects of floods (Kale, 1998) and by discovering the real nature of monsoon floods (Baker, 1994; Baker 1998). Hence, the expansion, improvement and modernization of the existing network of stream gauging stations, and the systematic monitoring of: (a) the climatic and hydrological systems, and (b) the geomorphic and environmental performance of the existing flood-protection works (Hey, 1994) are vital from the standpoint of flood hazard management (Kale, 1998).

Furthermore, it is also necessary to recognize the following aspects of floods, that have emerged from flood studies in India and other flood-prone countries, while undertaking all types of future flood control and management programs.

a) The occurrence of flood is unavoidable, since the phenomenon is an integral part of the monsoon climate. In view of the inherent high inter-annual as well as inter-regional variability in the monsoon rainfall, floods (as well as droughts) are inevitable. Flooding is, therefore, nothing unusual in the case of monsoonal rivers. Floods have been occurring on these rivers for tens of thousands of years, and will continue to occur in the future also.

b) On a long-time scale, most rivers exhibit rainfall-related changes in flood activity and rainfall-driven alternating periods of low and high flood activity. In general, there is a good correspondence between periods of increased average rainfall and increased frequency of high-magnitude floods. Therefore, these decadal shifts in the rainfall and flood regime conditions must be taken into consideration when attempting to estimate flood peaks or return periods for flood prediction and flood forecasting.

Floods constitute an integral element of the natural cycle of the flow of monsoonal as well as seasonal rivers that primarily determine the channel forms and processes, as well as the nature of aquatic life and riparian vegetation (Kale, 1998). It is now a well-established scientific fact that in monsoon regions floods are important hydrologic and geomorphic events (Kale, 1998; Ward, 1978). Therefore, heavy flood-control works would undoubtedly have an adverse affect on the fluvial as well as flood regime conditions, and this in turn would have a lasting effect on the riverine environment. While structural measures, such as dams, barrages and embankments have proved beneficial to some extent, their undesirable effects on siltation of rivers and bank erosion are huge. Further, the available information indicates that the impacts of floods resulting from the breaching of the flood barriers are much more severe than floods on rivers without any intervention. Therefore, heavy flood control works, particularly on unstable rivers, should be avoided, and the focus should be shifted for hard measures (structural) to soft measures (non-structural).

In the last few years, flood scientists all over the world have begun to realize that flood is a natural phenomenon, seemingly difficult to control or suppress. Several devastating floods in the last few years in many parts of the world (Bangladesh and India: 1987–1988; USA, Mississippi: 1993; Central Europe, Oder: 1997; Bangladesh, India and China: 1998; China and Southeast Asia: 1999, 2000, etc) have vividly demonstrated the limitations of the existing hydrological techniques in forecasting floods, and have also indicated that structural methods are not always viable solutions, especially in controlling rare, large floods.

As flood hazards are inevitable and will continue to occur, it has been suggested that the flood vulnerability in India and other flood-prone regions of the world, could be reduced by adopting more river-friendly schemes that are based on geomorphic principles and those that recognize the geomorphic importance and the environmental value of floods (Kale, 1998; Hey, 1994; Baker 1998).

ACKNOWLEDGEMENTS

This work was carried out in connection with three research projects supported by the Department of Science and Technology, New Delhi. The author would like to thank the officials of the Central Water Commission for providing hydrological data in connection with the above projects.

REFERENCES

Central Water Commission Publication (1996) *Water and Related Statistics,* New Delhi, p. 232.

Dhar, O N, Mooley, S S and Mandal, B N (1986) *Transactions Inst. Ind. Geog.*, **8**: 13.

Kale, V S (1997) *Jour. Geol. Soc. Ind.*, **49**: 359.

Kale, V S (1998) Flood Studies in India, In Kale, V S (Ed) *Geological Society of India, Bangalore Memoir*, **41**: 229.

Hofer, T (1998). Flood Studies in India, In Kale, V S (Ed) *Geological Society of India, Bangalore Memoir*, **41**: 119.

Sakthivadivel, R and Raghupathy, A (1978). *Hydrology Review,* **4**: 57.

Goswami, D C (1998). Flood Studies in India, In Kale, V S (Ed) *Geological Society of India, Bangalore Memoir*, **41**: 53.

Goswami, D C (1985) *Water Resources Research,* **21**: 959.

Seth, S M and Datta, B (1982) *Proceedings International Symposium on Hydrological Aspects of Mountainous Watershed.* University of Roorkee, Roorkee **III**: 1.

Baker, V R (1977) *Geol. Soc. Am. Bulletin.*, **88**: 1057.

McMahon, T A, Finlayson, B L, Haines, A T and Srikantah, R (1992) *Catena Paperback,* Cremlingen, p. 166.

Erskine, W D and Warner, R F (1988) In Warner, R F (Ed) *Fluvial Geomorphology of Australia,* Academic Press Sydney, p. 223.

Dhar, O N and Nandargi, S (1998) Flood studies in India, In Kale, V S (Ed) *Geological Society of India, Bangalore Memoir*, **41**: 53.

Khan, M (1969) *Floods and their Computation,* IAH-UNESCO-WHO Publication, Leningrad, p. 918.

CSE (1991) *Floods, Flood Plains and Environmental Myths.*, Center for Science and Environment, New Delhi.

Bhalme, H N and Jadhav, S K (1984) *Jour. Climatol.*, **4**: 509.

Parthasarathy, B, Rupa Kumar, K and Munot, A A (1991) *Jour. Climate,* **4**: 927.

Whetton, R and Rutherfurd, I (1994) *Climate Change,* **28**: 221.

Ramaswamy, C (1987) *Meteorological Aspects of Severe Floods in India 1923–1979*. India Meteorological Department, New Delhi, MMH No. **10**.

Rakhecha, P R and Pisharoty, P R (1996) *Current Science,* **71**: 179–186.

Nandargi, S S (1996) *Rainstorm Studies for Planning and Development of Water Resources of the Indian Region.* Unpublished Ph.D. Thesis, University of Pune, Pune.

Chaphekar, S B and Mhatre, G N (1985) *Human Impact on Ganga River Ecosystem,* Concept Publishing Company, New Delhi.

Messerli, B and Hofer, T (1995) *Water and the Quest for Sustainable Development in the Ganges Valley,* Chapman, G P and Thompson, M (Eds) Mansell, New York, p. 64.

Brammer, H (1990) *The Geog. Jour.,* **156**: 158.

Probst, J I and Tardy, Y (1987) *Jour. Hydrol.,* **94**: 289.

Burn, D H and Arnell, N W (1993) *Jour. Hydrol.,* **144**: 381.

Kripalani, R H and Kulkarni, A (1997) *Weather,* **52**: 39.

Kale, V S (1999) *Jour. Geol. Soc. Ind.,* **53**: 5.

Wright, P B (1989) *Int. Jour. Climatol,* **9**: 33.

Gregory, S (1989) *Int. Jour. Climatol,* **9**: 465.

Krishnamurthy, V and Goswami, B N (2000) *Jour. Climatol,* **13**: 579.

Baker, V R (1994). *Geomorphology,* **10**: 139.

Baker, V R, Ely, L L, Enzel, Y and Kale, V S (1995) Quaternary environments and geoarchaeology of India, In Wadia, S, Korisettar, R and Kale, V S (Eds) *Geological Society of India, Bangalore Memoir,* **32**: 61.

Hollander, M and Wolfe, D A (1973) *Nonparametric Statistical Methods*, John Wiley and Sons New York, p. 192.

Chiew, F and Mcmohan, T (1993) *Int. Jour. Climatol.,* **13**: 643.

Marengo, J A (1995) *Climate Change,* **31**: 99.

Imperial Gazetteer of India (1908) The Indian Empire, Oxford, Clarendon Press.

Ely, L L, Enzel, Y, Baker, V R, Kale, V S and Mishra, S (1996) *Geol. Soc. Am. Bulletin,* **108**: 1134.

Kale, V S (1999) *Man and Environment,* **24**: 109.

Kale, V S, Mishra, S and Baker V R (1997) *Jour. Geol. Soc. India,* **50**: 285.

Inglis, C C (1949) *The Behaviour and Control of Rivers and Canals,* Central Water Power Irrigation and Navigation Research Station, Pune, **13**: 90.

Chitale, S V (1978) *Irrigation and Power Jour.,* **36**: 493.

Hey, R D (1994) *Rivers Handbook II,* Callow, P and Petts, G (Eds) Blackwell, Oxford, p. 337.

Sharma, P K, Chopra, R, Verma, V K and Thomas, A (1996) *Int. Jour. Remote Sensing,* **17**: 3511.

Rodriguez, C R and Rao, U R (1993) *Acta Astronautica,* **30**: 93.

Baker, V R (1998) Flood Studies in India, In Kale, V S (Ed) *Geological Society of India, Bangalore Memoir,* **41**: 219.

Ward, R (1978) *Floods: A Geographical Perspective,* The MacMillan Press Ltd., London, p. 56.

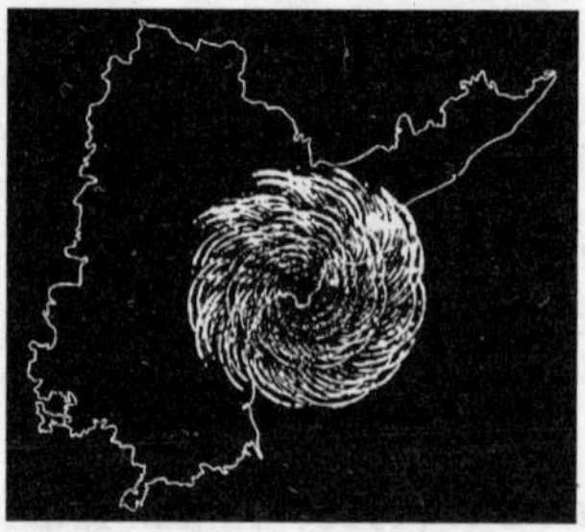

Monitoring and Forecasting of Tropical Cyclones and their Associated Effects

S R Kalsi*

Abstract

The India Meteorological Department (IMD) was instituted in 1875, mainly to cater to the requirements of Tropical Cyclone (TC) warning. Since then, we have come a long way in developing and improving infrastructural facilities in this connection. The monitoring process has been revolutionized by the advent of remote sensing techniques. Satellite image interpretation has done wonders in meeting the requirements of a TC intensity analysis and forecast. Numerical techniques are being applied for forecasting TCs and their effects. This paper reviews the progress made in IMD in this regard, and also some of the disaster management aspects of the TC warnings.

Introduction

Tropical cyclones (TCs) are intense, low-pressure systems that develop over warm seas. They are the most destructive atmospheric phenomena and are capable of destroying coastal cities and killing thousands of people due to strong winds, heavy rain, and storm surges. About 7% of the global crop of TCs develop over the North Indian Ocean. About 5 to 6 TCs form in the North Indian Ocean each year, with a variation of 1 to 10 during the period 1890 to 1999. The frequency of the TC in the Bay of Bengal is 4 to 5 times more than

* India Meteorological Department, New Delhi

that in the Arabian Sea. About 35% of the initial disturbances in the North Indian Ocean reach the TC stage, while 45% of these TCs become severe. Although large variability in the locations of TCs from decade to decade has been reported, no systematic pattern has emerged. The seasonal variation has a bimodal distribution, with the primary maximum in November and the secondary maximum in May. The intervening period of the summer monsoons is relatively free of TCs as the monsoon trough, which is usually the seat for cyclogenesis, shifts to the north. In the Gangetic plains, the eastern end of the monsoon trough dips into the head of the Bay of Bengal. The monsoon depressions form in the cyclonic shear zone located over head of Bay. These are benevolent rain-bearing systems and cause heavy rains over central and northern India along their track. Since these depressions form close to the coast, they do not generally intensify to the stage of TCs, although occasionally a few of them may acquire the characteristics of a marginal cyclone. When the monsoon trough is located further south over the central and southern Bay of Bengal, these depressions become TCs before the landfall. Although the development from a tropical depression into a TC usually occurs in 12 to 24 hours, 15% required more than 48 hours, and other are reported to undergo formation in less than 12 hours. Formation may occur either as re-intensification of westward-moving disturbances, or from the disturbances that develop within the trough.

In this paper, the observational resources that are available, or will become available in the near future, for analysis and prediction of TCs are described in Section 2. Section 3 describes the structural evolution of the TCs, whereas Section 4 is devoted to techniques dealing with forecasting the motions of TC. The important aspects of strong winds, heavy rainfall and storm surges are dealt with in Section 5, and disaster management aspects of TC warnings are included in Section 6. Weaknesses in the warning system are summarized in Section 7 and future scenarios figure in Section 8, with a listing of the gaps in the system and recommended actions for improvement.

Data Resources

TCs are characterized by two predominant scales of atmospheric motion: large scale (1000 km) and TC scale (100 km). We need to have various observational systems to resolve the scales associated with the TC vortex, its surrounding bands, and the large-scale environmental flow in which the TC is embedded. For an appropriate appreciation of the TC prediction problem, we require a thorough knowledge of the availability and distribution of the existing data resources. Although satellite-based observations are being extensively utilized now, the use of aircraft reconnaissance and Rawin sonde have generally decreased in the tropics. The distribution of data is largely an economic function and varies considerably among the TC basins.

Surface

The IMD operates a surface network of over 550 stations, but TCs have been analyzed with data mainly from ships of opportunity. However, this data cannot result in an accurate positioning of the TC center. At the same time, there has been considerable decrease in ship observations from the storm field area, due to advanced warnings available to ships to avoid the TC regions altogether. Ships now supplement the standard land-based measurements. Instrumented ocean research vessels have also been available. Their numbers are also decreasing because of the high expenses involved and they are being slowly phased out, even in higher latitudes of the northern hemisphere, where they have operated on station for many years. Satellite-integrated automatic weather stations have also been installed on islands, oil-rigs, and exposed coastal sites. Drifting buoys, supplementing the surface data network in the tropical ocean, have also been deployed. Drifting buoys have been deployed elsewhere from aircraft in advance of a hurricane (Black *et al.*, 1988) and successfully reported surface pressure, wind speed at a 1-meter level, and ocean temperature at sea levels. The Department of Ocean Development, Govt. of India, has also started a very important national data buoy program. They have since deployed a set of 12 moored buoys in the northern Indian Ocean, to provide meteorological and oceanographic data. This data network is going to be augmented in the next couple of years. Observations on storm surges have been a relatively neglected aspect; these observations are important to validate the storm model in all cyclone basins.

Upper Air

Dynamical forecasting of TCs requires knowledge of the vertical structure of both the TC and the surrounding environment. The rawin sonde (RS) remains the principle equipment for sounding. The availability of upper-level data is marginally less than with the surface data and may be reduced further as the cost of expendable (sondes, target and gas) impacts on the budgets of many developing countries that have an upper-air networks over the tropics. Alternatively, relatively expensive communication systems are available. The IMD operates a network of 65 pilot balloon observatories and 35 RS stations to meet these requirement. A mesosphere, stratosphere, troposphere (MST) radar has also been installed at Tirupathi, in India. Another profiler is being developed by Sameer and will be deployed at the IMD, Pune. Another important source of upper level data is the aircrafts' reports (AIREPS). An increasing number of commercial jet aircrafts are equipped with the Aircraft Meteorological Data Relay (AMDAR) system. This data is being made available for the forecasters and also being used by the IMD analysis and predictions scheme.

Aircraft Reconnaissance

Much of our knowledge about the inner region of the TC has been derived from research and reconnaissance aircrafts. They have been providing valuable information about the storm locations and intensity. We have also learnt a lot about the structure of the TC, using data from these flights, but this is a very expensive exercise and the USA has withdrawn this facility for probing TCs in the North-western Pacific in 1987. Remote-piloted, long-endurance aircraft have also been proposed. Holland *et al.* (1992) had flight-tested a much smaller aircraft equipped with meteorological sensors. This will provide valuable observations and the cost would be comparable to that of the rawin sonde, if the projected production costs are maintained.

Radar Observations

Radar has been used to observe TCs ever since the mid-1940s, from different platforms with varying degrees of sophistication. Surveillance of the spiral rain-bands and the eye of the TC is an important function of the coastal radars. We have installed a group of 10 Cyclone Detection Radars (CDRs), shown as per Figure 1. These radars are providing useful estimates of storm centers up to a range of 300–400 km. Raghavan (1997) gave an exhaustive review of radar observations of TCs in the Indian seas. It has been used in operational TC tracking and forecasting and also to understand the structure of TCs. It has also enabled us to have an estimate of the recent 6–12 hour motion of the storm, which is a key predictor in short-term track prediction. Observations of concentric eyewalls in an intense storm (Willoughby *et al.,* 1982) have been very important from the theoretical and operational point of view, but the radar data have not been of much use in the assessment of the intensity of TCs. Same difficulties have been experienced in tracking recent, very severe TCs in the Arabian Sea, wherein it was found that deployment of a CDR at Veraval would have been more effective in comparison to a CDR positioned at Bhuj. Doppler radars are being used the world over in connection with monitoring of TCs. The IMD operates a variety of radars, including systems which are used for wind finding and weather monitoring. They are operating in S-Band and X-Band. The ten old X-band radars have been upgraded into digital radars. Bhatia *et al.* (1999) have described the capability and applications of derived products and

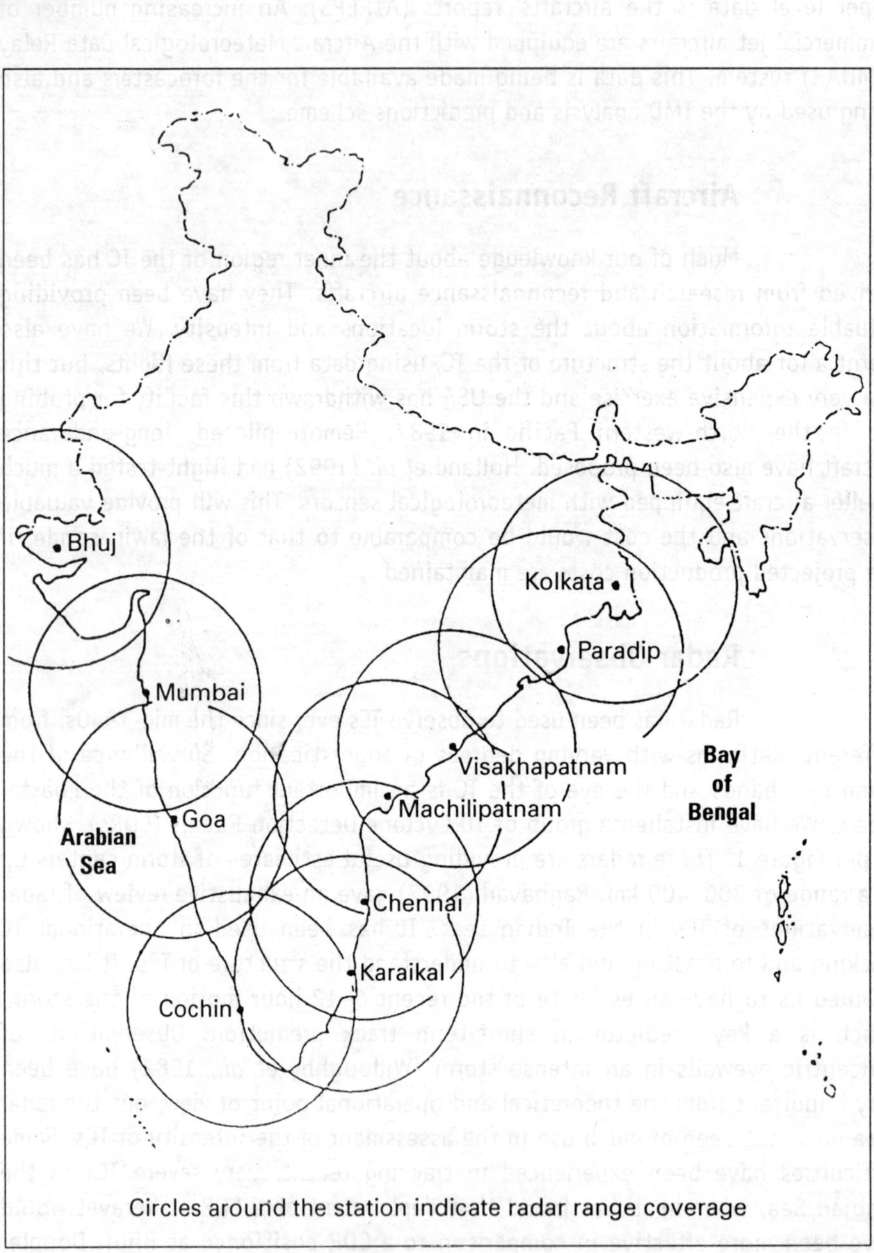

Circles around the station indicate radar range coverage

Figure 1 *Indian meteorological department cyclone detection radar network; effective range—400 km.*

also the plans for the deployment of Doppler radars that are providing direct measurements of wind fields in TCs. Due to the range limitation, Doppler wind estimates are usually within a range of about 100 km. The IMD is deploying Doppler radars at 3 sites in the east coast in the near future. Another set of 3 Doppler radars would follow thereafter, under the World Bank-aided scheme for Andhra Pradesh. Radars have also been flown on board satellite systems. One such system, launched in the recent past, is the Tropical Rain Measuring System (TRMM), which is described below.

Satellite Observations

The meteorological satellite has made a tremendous impact on the analysis of TCs. All developing cloud clusters are routinely observed through satellite cloud imagery, and those showing signs of organization are closely monitored for intensification. TC forecasters everywhere use the Dvorak (1984) technique to estimate storm location and intensity. It has been found to provide realistic estimates, even for TCs in the Bay of Bengal as well the Arabian Sea (Kalsi, 1992; Joshi *et al.*, 1999). INSAT data has also been used to study the structures of different TCs in the Bay of Bengal (Kalsi *et al.*, 1996). The IMD is also producing Cloud Motion Vectors (CMVs) at 00,06 and 12 GMT, but unfortunately low level CMVs are not available for TCs. The Very High Resolution Radiometer (VHRR) payload on board the INSAT-2E was improved upon to provide water vapor channel data in addition to VIS and IR onboard INSAT-2E. A separate payload known as the Charged Couple Device (CCD) has also been deployed on board this satellite. But VHRR has encountered anomalous behavior, which is currently being looked into by the Department of Space. Efforts are also underway to build a geostationary satellite purely for meteorological purposes, which will have capabilities both for sounding and imaging missions by a 2003 time-frame.

Infrared data from geostationary satellite is being used to estimate convective precipitation using the Arkin (1983) method. Water vapor winds from METEOSAT-V which has been deployed along 63° E longitude are also available over the Indian Ocean, but very important observations from satellites such as the Defense Meteorological Program Satellite (DMSP) and the Earth Research Satellite (ERS) are still not available on real-time bases to TC forecasters in India. Attempts have also been made in India for the extraction of useful information from the data provided by these satellites. Joshi *et al.* (1999) have reviewed the utility of the data from meteorological satellites, including the TRMM and ERS for the analysis and forecasting of TCs. Velden *et al.* (1989) had described the advantages of center-fixing at 85 GHz imagery compared with the conventional visible (VIS) and Infrared (IR), where the centers are obscured by cirrus overcast. The data from these new satellites is, therefore, very important from an operational point of view.

The forecaster rarely has sufficient observations to exactly define the structure and position of TCs and the adjoining synoptic systems. Further, in view of the new developments in marine communication through the Global Marine Disaster and Safety Systems (GMDSS), there has been a considerable loss of ship observations, particularly from the Bay of Bengal, after the closure of coastal radio stations operated by Department of Telecommunication (DOT), as the ship data collection using the GMDSS system has not been operationalized as yet.

Structure of Tropical Cyclones

The inner region, termed as the cyclone core, contains the spiral bands of precipitation, the eyewall, and the eye that characterize TCs in the radar and satellite imagery (Dvorak, 1975). The inner region winds can become intense and, in extreme cases, reach 90 m/s just outside the eye. The primary circulation – tangential or swirling wind in the core – is strongly axis-symmetric. TC structure forecasting has received considerably less attention than has motion forecasting because this is a more difficult problem, as the dynamics are more highly non-linear and occur on smaller scales under conditions that make them very difficult to observe. Prediction involves the analysis of present conditions and estimation of change, i.e., monitoring and forecasting. In the absence of most of the conventional and non-conventional data resources, emphasis is laid on inferring location, intensity, size, and rainfall from frequent satellite imagery.

Satellite Image Interpretation

The Dvorak technique has been calibrated using reconnaissance aircraft data in the North Atlantic and Western Pacific. It has also been used operationally for TCs in the Bay of Bengal (Kalsi, 1992). It is very well known that this technique fails miserably to account for explosive developments. Though rapid deepening rates of more than 42 hPa per day estimated with the help of satellite image analysis have been seen to be realistic in some cases, problems have been experienced while analyzing rapidly developing TCs in the Bay of Bengal in which the intensity increased by more than T 2.0 in the pre-hurricane stages of evolution, which is not allowed for by the Dvorak algorithm.

As far as the Bay of Bengal TCs are concerned, it is seen that there may not be a perceptible difference between developing and non-developing mesoscale cloud clusters in the beginning (Kalsi, 1993). However, the genesis phase is revealed by relatively improved organization of random mesoscale convection. A large-scale convective surge associated with the large cyclonic circulation may also be seen to precede the genesis. The persistence of the convection as a prelude to cyclogenesis was a feature common to many Bay of Bengal TCs. Kalsi and Jain (1989) also showed that the TCs developing in highly-sheared flows in the month of October, finish as marginal cyclones with wind speed reaching upto 55 knots. The hurricane stage termed as very severe cyclonic storm stage is associated with a banding eye pattern, in which a curved band spirals fully around the center of the cloud system in the VIS imagery. A sharp, circular and relatively dark eye appears in the Bay of Bengal TCs at sustained wind speeds of about 90 knots, about 12 hours later in IR imagery in most TCs.

The recent experience of the absence of the eye in the IR imagery, even at a wind speed of about 100 knots in the Kandla TC of 9 June 1998, the Gopalpur TC of 17 October 1999 and the Paradip Super TC of 29 October 1999 (at the stage of intensity of 100 knots) shows the enormity of the problem involved in using INSAT imagery to provide TC warnings. After the appearance of the eye, the intensity has to be upgraded very fast in the most of the TCs in the North Indian Ocean. Even the Joint Typhoon Warning Centre (JTWC) of the United States of America has showed rapid and dynamic intensification in some cases for the Bay of Bengal TCs in their operational advisory bulletins. The relevance in the operational warning of TCs of these enhanced estimates of intensity for storm-surge prediction becomes limited, as the intensification takes place within 24 hours of landfall and it is not very well accounted for about 2 days earlier. Though the inner edge of the eyewall is usually clear in the satellite imagery at this stage in the North Indian Ocean, the outer edge seems to be submerged in the Central Dense Overcast (CDO) that develops because of the left-over condensate, which is distributed horizontally in the upper troposphere by the outflow. Even the second eyewall seen in some intense TCs has not been resolved properly in most IR and some VIS images. These outer eyewalls are submerged in the CDO, though the moat between them is clearly seen in some images (Plate 1).

Observing systems, such as satellites and radars, have enabled us to monitor many facts of evolutions of intense TCs. The weakening of the momentum field associated with tangential winds has been preceded by the weakening of the convective structure of TCs. Satellites have enabled us to understand cases of shearing of TCs on account of interactions with strong westerly jets. Tropical convective systems, especially those at sea, undergo large diurnal variation because of the different in short-wave and long-wave radiation budgets of clear and cloudy regions (McBride and Gray, 1980). Kalsi *et al.* (1996) have documented the structural variability of TCs in the Bay of Bengal. In a particular case, they have inferred rapid deepening with pressure falls of more than 42 hPa, on 7 May 1990. This cyclone also displayed double eyewalls both in satellite and radar imagery (Kalsi *et al.*, 1993).

Forecasting Structure Changes

This is one of the most difficult problems. At present, it is done by using the Dvorak model of development, statistical methods, and subjective assessment based on satellite advisory and environmental conditions. Some kind of checklists have been devised to work to the intensity changes. The favorable and unfavorable signatures are discerned from satellite imagery. The vertical shear is calculated from the average winds in the upper and lower troposphere. The position and movement of the adjacent synoptic systems and westerly troughs are also discerned from satellite imagery. All these factors taken together give an indication of whether or not the disturbance is going to intensify further. The intensity change at present is not very well captured in the numerical forecast models.

The outer-core structure forecasting has not received as much attention as the inner-core structure, but it is important as it determines the vulnerability of the persons and property at risk. No standard definition exists for out circulation; however, it may be seen as the extent of the gale force winds at the surface. Another measure is the radius of the outer closed isobar. Both measures are poorly related to intensity (Weatherford and Gray, 1988). Filling cyclones are seen to have a larger and a stronger outer-core circulation, which can be easily observed by ships which usually skirt the inner-core of cyclones. Though the inner core intensity is inferred using satellite imagery, the outer core is more difficult to infer in this way. However, data from passive microwave radiometers could be of great use for observing outer core circulation. Some rules have been developed to visualize the outer core circulation and its changes. Weatherford and Gray (1988) drew a lot of useful inferences from the point of view of forecasting of TCs and their attributes.

Motion of Tropical Cyclones

The TC track-prediction technique can be broadly divided into two categories: i) subjective, and ii) deterministic. Synoptic, Satellite, Radar Method, etc., come under subjective techniques. Statistical and dynamical methods are broadly categorized as objective or deterministic techniques. Figure 2, reproduced from Mohanty and Gupta (1997), shows different methods of track prediction under these categories. Some of these are described in the following paragraphs.

Forecasting the movement of TCs is universally considered to be the most important task. A large number of statistical/empirical methods have been developed in the past and, despite the numerical models which have been commissioned in most of TC forecast offices, they are still being used and

Plate 1: Satellite image of a very severe cyclonic storm in the Arabian Seas showing multiple eyewalls.

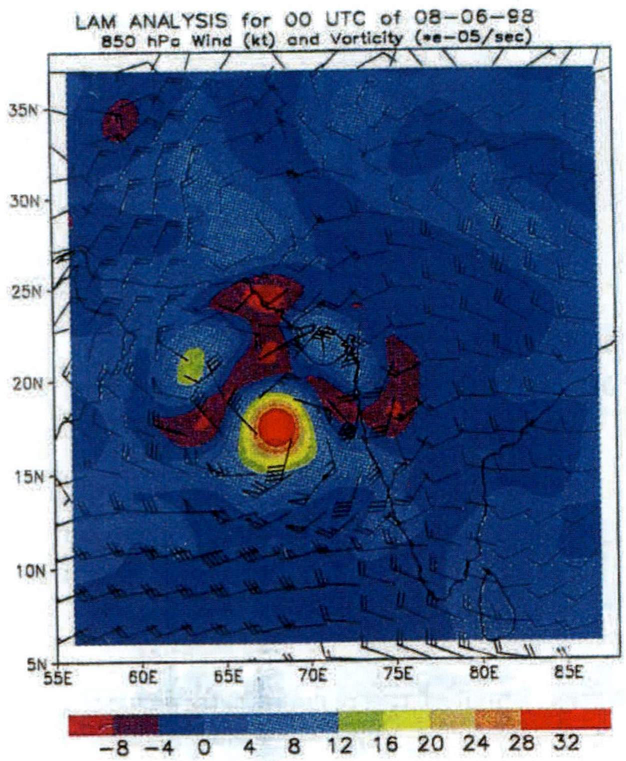

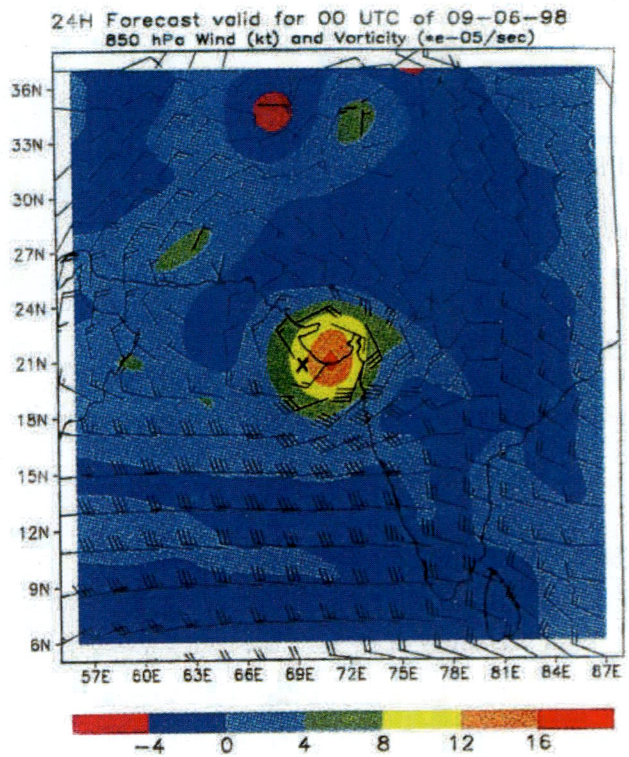

Plate 2 'X' represents the realized position of the vorticity center at 00 UTC, of 9 June 1998.

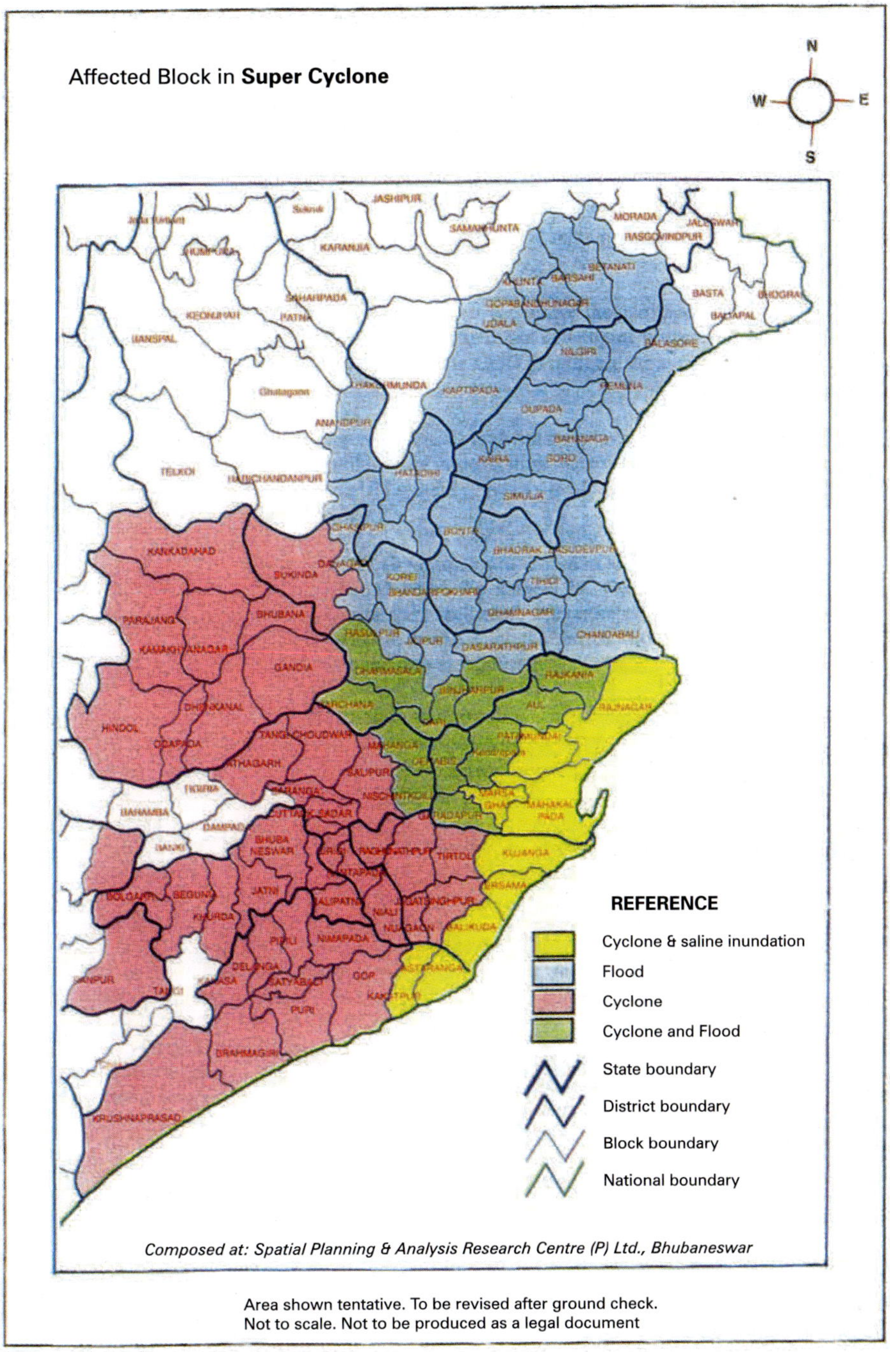

Plate 3. Impact of the Super Cyclonic Storm in Orissa.

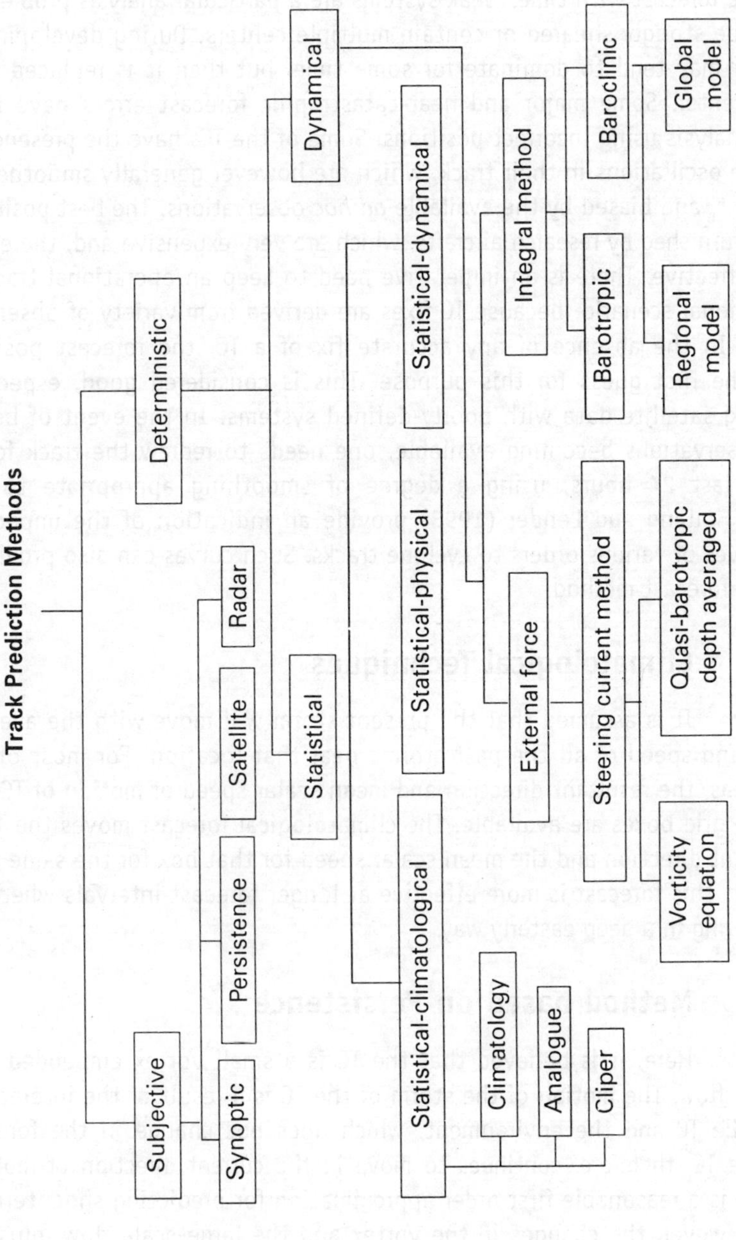

Figure 2 *Methods of tropical cyclone track prediction.*

improved upon. The first step in making a TC track forecast involves fixing its position using all available data resources. Highly accurate positioning is important for accurate short-term forecast. The initial error in the location of TCs vitiates the forecast with time. Weak systems are a particular analysis problem as they may be stongly sheared or contain multiple centers. During development, one center may tend to dominate for some time, but then it is replaced by a separate center. Some major and near-catastrophic forecast errors have been made by analysis using incorrect positions. Some of the TCs have the presence of small-scale oscillations in their track, which are however generally smoothed to some extent, and biased by the available *ad hoc* observations. The best positions of TCs are furnished by research aircrafts, which are very expensive and, therefore, not cost effective. There is an imperative need to keep an operational track in the operational scenario, because TC fixes are derived from variety of observing platforms. In the absence of any accurate fix of a TC, the forecast position provides the first guess for this purpose. This is considered good, especially when using satellite data with poorly-defined systems. In the event of better quality observations becoming available, one needs to redraw the track for at least the last 24 hours, using a degree of smoothing appropriate to the situation. Holland and Lender (1993) provide an indication of the impact of fitting curves of various orders to cyclone tracks. Such curves can also provide a short-term forecast method.

Climatological Techniques

It is assumed that the present storm will move with the average direction and speed of all the past storms near that location. For most of the motion areas, the resultant direction and mean scalar speed of motion of TCs for 2.5 degree grid boxes are available. The climatological forecast moves the TC in the resultant direction and the mean scalar speed for that box for the same time of the year. This forecast is more effective at longer forecast intervals when the TCs are moving in a deep easterly way.

Method based on Persistence

Here, it is believed that the TC is a small vortex embedded in a large-scale flow. The motion of the storm of the TC is a result of the interaction between the TC and the environment, which does not change in the forecast period. The TC, therefore, continues to move in the current direction of motion. Persistence is a reasonable first order approximation for predicting short-term TC motion. However, the changes in the vortex and the large-scale flow, plus the inherent non-linearity of the interaction between them, are the limitations of this method.

Persistence and Climatology

These methods are based on the assumption that the integrated efforts of all forces, which have steered the TC during some past period, will continue to be effective during some future period. The persistence forecast is the linear extrapolation over the next 12 hours of the smoothed path of the storm, in the past 12 to 24 hours. This is a simple method. The temporal and spatial repetitiveness of the cyclone tracks, produced by the synoptic pattern in steering the TCs, is helpful in making predictions based on climatology.

Steering

This method is based on the assumption that the storm is steered by the basic current in which it is embedded. Options are strongly divided in choosing the levels which steer the cyclonic storm. Chan and Gray (1982) presented a comprehensive study of the relationship between the movement of TCs and the large-scale circulation in which they are embedded.

The relationship between the TC and its environment has been examined by several authors. George and Gray (1976), and Chan and Gray (1982) estimated the basic current by averaging a 1 to 7 degree latitude radial band surrounding the TC and found significant and consistent track deviation. In the mean, the TC tends to move poleward and westward of this basic current. For example, low-latitude, westward-tracking TCs move faster and slightly poleward of the basic current, those moving to the northeast move slower and to the west. It is also seen that the very large TCs tend to move more independently of the environmental flow than do the small systems. The appropriate layer or layer for steering has also been debated. The most consistent observational result is that 700 or 500 hPa level provides the closest approximation. Holland (1984) has argued against the inclusion of inflow and outflow layers and recommended the layer from 850–300 hPa.

Synoptic Method

The most widespread synoptic pattern recognition involves the relative locations of the cyclone with the sub-tropical ridge and upper-level trough in the mid-latitude westerlies. Generally, though not always, a strong sub-tropical ridge located ahead of the cyclone will result in a continued westward movement. Similarly, the presence of a westerly trough in the upper level, west of the TC, is an excellent indicator of recurvature.

At a smaller scale, interactions between adjacent weather systems can result in a consistent track pattern that has a high degree of predictability, provided that the interactions are recognized. The most famous of these is the interaction of binary vortices, known as the "Fujiwhara Effect".

Dynamical Methods

Dynamical methods have proved successful because of the increase in the power of computing, the development in numerical prediction techniques, the improved understanding of physical processes, improvements in observing systems, and objective analysis and initialization. Considerable success has been achieved using dynamical models (Dastoor and Krishnamurty, 1991; Mathur, 1991). Some efforts have been made by Indian workers towards forecasting the movement of depressions and cyclones (Ramanathan and Bansal, 1977; Sikka, 1975; Mohanty, 1994). The problem continues in the tropical prediction for want of adequate and accurate data to define the initial conditions. To overcome this problem, many workers together constructed an artificial vortex and merged it with the large-scale analysis (Mathur, 1991; Iwasaki *et al.*, 1987; Singh *et al.*, 1990). A number of techniques are in vogue the world over, but only the operational models in use in India are discussed below.

A global spectral model (T-80/L-16) and a global data-assimilation system based on short range forecast (six hours) of this model, involving the Statistical Spectral Interpolation (SSI) scheme of analysis, has been operational at the NCMRWF since June 1994. The satellite data sets (SATEM and SATOB) along with a variety of other conventional and non-conventional data, received on the GTS or through the Internet, are routinely being used on an operational basis. Gupta and Romesh (1999) have evaluated the performance of this model in predicting tropical cyclogenesis over the north Indian Ocean. Due to the paucity of observations over the oceanic areas, the problem of non-capturing of TCs in numerical analysis still remains unresolved.

Prasad *et al.* (1995) have discussed the details of a Limited Area Analysis and Forecast System (LAFS), run operationally by the IMD. It consists of an automated data-decoding procedure, a multivariate optimum interpolation scheme of objective analysis, and a high-resolution (1° × 1° lat/long grid) multi-layer (12 sigma levels) primitive equation model. It runs in operational mode on a Cyber 2000U computer system in IMD, and provides data twice a day based on 0000 UTC and 1200 UTC inputs. The analysis system uses observational data from the surface (SYNOP/SHIP), the upper air (TEMP/PILOT), satellites (SATEM/SATOB) and aircrafts (AIREP), which are extracted and decoded from the raw GTS data sets. All the data are quality controlled and packed into a special format for objective analysis. The first guess and lateral boundary data files are constructed from the global model forecasts run by NCMRWF. The LAFS gives several products of basic flow variables and derived fields, as well as forecasts of accumulated precipitation. The forecast model is a semi-implicit, semi-Langarangian, multi-layer, primitive equation model cast in the sigma coordinate system and the staggered Arakawa C-grid in the horizontal (Krishnamurthy *et al.*, 1990). The present version of the model has a resolution of 1° × 1° lat/long. (91 × 51 grid; covering a domain 10°S–40°N and 40°–130°E) in the horizontal and 12 sigma levels (1.0 to 0.05) in the vertical. This model incorporates physical processes such as planetary boundary-layer parameterization, shallow and deep cumulus convection, and radiation etc. Enveloporography I, used in the current version model, is run up to 48 hours forecasting.

Prasad (1997) described the synthetic vortex-generation scheme for numerical forecasting of TCs. The scheme basically generates radial distribution of surface pressure within the vortex, from an empirical formula proposed by Holland (1980). The basic inputs for generating the surface pressure are the parameters i.e., the central pressure of the storm, its environmental pressure, radius of the maximum wind, the current position, movement and intensity of the storm, which are inferred from the surface synoptic charts and satellite imagery. Once the surface pressure field has been constructed, the surface winds are obtained from the gradient-wind relationship. The upper winds are computed from the surface winds by using the composite, vertical, wind-shear factors proposed by Anderson and Hollingsworth (1988).

Plate 2 shows the actual and predicted wind and vorticity fields on 8 and 9 June 1998 for the Kandla TCs. Kalsi (1999) showed that the performance of the LAM model in vogue at New Delhi is comparable to that of other similar models deployed elsewhere the world over, especially in view of the fact that there are problems in tracking TCs in the north Indian Ocean (Elsberry, 1998). It was shown in Prasad *et al.* (2000) that the mean track error was 169 km for 24 hours and 254 km for 48 hours, for the 1998 cyclone season of the north Indian Ocean. The performance was comparable to the UKMO global model for that year.

Recently, a quasi-Lagrangian Model (QLM) for cyclone track prediction in the Indian seas has been implemented in the IMD. The model is an adapted version of the hurricane prediction model of the National Center for Environmental Predictions (NCEP—the erstwhile National Meteorological Center) Washington (Mathur, 1991). It is a multi-level primitive equation, a fine-mesh model cast in the sigma coordinate system (p = p/ps; pressure divided by surface pressure). The model has a limited area domain, using a Cartesian grid. The horizontal grid spacing is 40 km and the integration domain consists of 111 × 111 grid points in a 4400 × 4400 km2 domain that is centered on the initial position of the cyclone. The QLM uses 16 laryers in the vertical. The model incorporates physical processes, which include surface frictional effects, sea-air exchange of sensible and latent heat, convective release of latent heat, divergence damping, horizontal diffusion, and isobaric condensation of water vapor. Radiation and turbulent processes, which only have a marginal impact on the development, are currently excluded to minimize computational time. The numerical integration of the model is carried out by using the so called quasi-Lagrangian method.

The model provides track forecasts out to 36 hours at present. The initial analysis and lateral boundary conditions are generated from operational analysis and forecasts produced by the global spectral model of NCMRWF, New Delhi. The initial fields for QLM are obtained as follows. First, the analysis valid at the map time is carried out by updating the NCMRWF GCM forecast, 12H (for 12 UTC run) or 24H (for 00 UTC run) with current observations by the Optimum Interpolation (OI) scheme. The NCMRWF forecast fields are a set of spectral coefficients, being the outputs of a T80 GCM on 18 sigma levels. The spectral coefficients are transformed to the QLM grid and vertical interpolation is carried out to get QLM sigma fields from GCM sigma levels. OI analysis is carried out directly on the QLM sigma levels. A new version of the IMD's operational optimum interpolation scheme for objective analysis (used for generating initial fields for IMD LAM) has been developed to suit the QLM grid structure, which is quite different from the grid structure of the LAM in both the horizontal and vertical. The symmetric vortex as described in the preceding subsection, and the analysis are then merged using appropriate weighting functions. The symmetric vortex fields are first projected on the QLM grid and then merged with the analyzed fields.

Prescription of a Steering Current

A steering current, which is specified based on the current storm speed and directions, is superimposed on the analyzed fields. The steering current is computed by constructing a dipole circulation. The dipole winds and geopotential height fields (incremental heights calculated from dipole winds geostrophically) are added to the vortex fields at all levels. Thus, the two special attributes of the QLM are: (i) the merging of an idealized vortex into the initial analysis, to represent a storm in the QLM initial state; and (ii) the imposition of a steering current over the vortex area with the use of a dipole. The full details of the model can be seen in Mathur (1991).

Track Forecast Experiments and Evaluation of Model Performance

The QLM is being run in an experimental mode since 1998, along with the LAM Model. Track forecast experiments were carried out with respect to cyclonic storms that formed during the three-year period from 1998 to 2000 in the Bay of Bengal and the Arabian Sea (Kalsi, 2001). A quantitative assessment of the performance of the forecast model can be made by the computation of track prediction errors. Two types of prediction errors have been attempted. Direct position errors (DPE) have been calculated by taking the geographical distance between the predicted position in each case of forecast and the corresponding observed position. The second type of error is the angular deviation between the observed and predicted track vectors, starting from a given initial position of the storm. While the former gives a measure of the absolute error of prediction, the latter provides an indication of the closeness of the predicted direction of movement and the observed direction.

As given in Kalsi (2001), Table 1 contains the verification statistics of the mean position errors (km) and the angular deviation of the predicted track from the observed track (degree), with respect to each of the seven cases studied. The mean position errors for a 24 hours forecast ranges between less than 100 km and a maximum of about 150 km. The 36 hour forecasts have these errors, varying between around 133 km to as much as 426 km. The angular deviations vary between about 5° to 30°. The overall average position errors for all the cases taken together (shown at the bottom of the table) work out to 123 km (24 hours), and 227 km (36 hours) and an angular deviation of less than 20 degrees for both hours.

Table 1 *Track prediction errors*

Year	Period	24 H		36 H	
		Mean position error (km)	**Angular deviation between observed and predicted track vectors@ (rmse) (deg)**	**Mean position error (km)**	**Angular deviation between observed and predicted track vectors@ (rmse) (deg)**
1998	20–22 Nov.	146.2	17.4	242.6	12.7
	6–10 June	140.1	32.4	205.2	27.9
1999	16–20 May	97.4	12.1	133.0	12.2
	15–18 Oct.	151.4	13.7	426.5	20.7
	26–30 Oct.	107.7	19.2	185.7	23.0
2000	15–19 Oct.	105.0	25.8	237.1	23.8
	27–30 Nov.	114.8	05.7	161.5	15.5
	25–29 Dec.	80.4	12.3	119.3	11.0
	Mean	**117.9**	**17.3**	**213.9**	**18.4**

@ Observed track vector: Initial (at T_0) to observed (at T_0+24H or T_0+36H) positions
Predicted track vector: Initial (at T_0) to predicted (at T_0+24H or T_0+36H) positions

In order to make a comparison of the QLM with some other cyclone-track prediction models run by various NWP centers, for which statistics were readily available, we present in Table 2 (a) and (b) the data of mean position errors with respect to some models run by China, Korea and UKMO, as well as the operational FSU-based, limited area model being run in the Indian Meteorological Department. It may be seen that the forecast errors of QLM are, in general, less than or comparable to all other models.

Table 2 (a) *Mean position (km) errors of cyclone track prediction by various NWP models (24 H)*

	IMD (LAM)	IMD (QLM)	China	Korea (1)	(2)	(3)	(4)	UKMO
1996			191					
1997			154	192	159	180	213	
1998	173(4)		143(2)					200
1999	186 (3)		119 (3)					123
2000			100 (3)					147

Table 2 (b) *Mean position (km) errors of cyclone track prediction by various NWP models (48 H)*

	IMD (LAM)	IMD (QLM) (36 H)	China	Korea (1)	(2)	(3)	(4)	UKMO
1996			356					
1997			321	467	294	361	307	
1998	236 (4)		224 (2)					310
1999	275 (3)		248 (3)					261
2000			173 (3)					245

LAM: Limited Area Model; QLM: Quasi-Lagrangian Model;
Korea: (1) Korea Typhoon Model (2) Geophysical Fluid Dynamics Model;
(3) Barotropic Adaptive Typhoon System (4) Global Data Assimilation System.

Winds, Rains and Storm Surges

Among all natural disasters, TC-caused disasters are always the worst both in terms of the death toll and the economic losses. These disasters are mainly caused by strong winds, torrential rains and storm surges. A mature

TC may have a wind speed of 60–70 m/s. This is the main driving force for the generation of storm surges, which incidentally depend on the offshore bathemetry. Although a TC usually reduces in intensity after the landfall, an enhanced supply of water vapor and interaction with moderately cold air may re-intensify the TC and release heavy rainfall. A TC does not have to be intense to produce very heavy rainfall; it all depends on the speed and size of the TC. The rainfall caused by TCs has been studied by various workers. Mukherjee *et al.* (1981) have found that the maximum amount of rainfall occurs on the western side, near the track of the TC. An analysis of the 30 years rainfall data of heavy rainfall associated with the TCs, indicate that rainfall of 30–40 cm has occurred in the case of severe TCs during the post-monsoon season.

Pattern recognition techniques are used not only for the movement of TCs, but also for forecasting their other attributes such as strong winds and heavy rains. Satellite and radar images too go a long way in meeting these requirements. Eye-walls and spiral bands are areas of intense weather and are the most conspicuous features in the satellite and radar images. Attempts are being made to simulate these features and also the severe weather associated with them, using the high-resolution, limited-area model. The amount of 24 hour rainfall is under-predicted (Sarkar *et al.,* 1999). In another study, Rai Bhowmik and Prasad (2000) also found some correspondence between the observed and the predicted rainfall fields, though dissimilarities were also noticed. Intensity forecasting continues to pose a challenge to the modelers. Hence, the problem in forecasting of storm surges is that they are usually encountered at a distance of about 50 kms from the point of landfall. Numerical studies of storm surges were started in India by Das (1972). Das *et al.* (1974) developed a storm surge prediction model for the north Bay of Bengal. This was followed by Ghosh (1977 & 1983) and also by an IIT New Delhi group led by Prof Dube, reviewed in Dube *et al.* (1997). The meteorological factors that determine the storm surge are the pressure defect at the center, the radius of the maximum winds along with the track and the speed of movement of the TC. The pressure defect is calculated using Mishra and Gupta (1976). The radius of maximum winds is derived from satellite and radar images.

In order to generalize the disaster that usually occurs in association with intense TCs, the case of the very severe cyclonic storm of 9 June 1998 is referred to here. The success and failure of early warning systems was discussed in this case by Kalsi and Gupta (1998). This is the first cyclone that has crossed the Gulf of Kutch and inflicted disaster in the Kandla area, which has never been witnessed there in the recorded history of cyclones. Earlier, a cyclone with a hurricane core of winds affected this area on 12 June 1964 (Raman, 1965) and produced storm tides of about 2 metres in height at Kandla. The overall damage and casualities encountered in that cyclone in which wind speed of 73 knots was recorded at Naliya, and the Porbandar cyclone of October 1975 (Mukherjee *et al.,* 1977) in which an easterly gust of 100 knots was reported from Jamnagar, were less in comparison. Figure 3 indicates the recorded or estimated wind speed encountered in Gujarat and Rajasthan, in association with this cyclone. It is interesting to find that Jamnagar and Kandla again report very strong winds (over 160 kmph in this case), stronger than the winds at Porbandar—the point of landfall. Again, the lowest surface pressure of 961 hPa is reported from Jamnagar. The October 1975 cyclone at Porbander showed a well-defined eye, which provided a realistic wind estimate of about 115 knots, corresponding to T6 intensity (Gupta *et al.,* 1977). The estimated or observed winds in the current cyclone seem to be stronger over Saurashtra and Kutch in comparison to

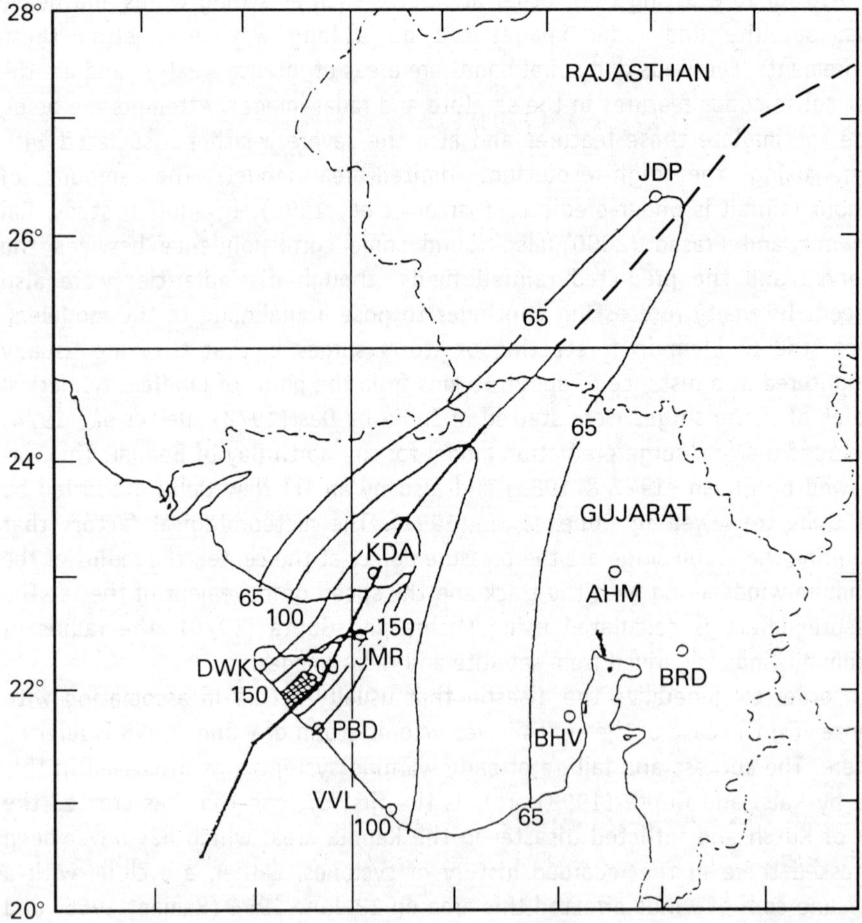

Figure 3 *Damaging winds during the June 1998 cyclone (km/hr).*

that of the earlier cyclone, but an exact comparison of the intensities of the two systems cannot be made in the absence of any reliable wind data. The diameter of the cyclone at the time of landfall seems to be very small, as Porbandar did not experience "eye" conditions. The eye, however, passed through Kandla as seen from the observations received from Kandla Aerodrome Observatory, which provided very useful hourly current weather observation. Very strong winds in excess of 180 kmph at Jamnagar and continuous rain or drizzle there indicate that it was not on the track of the eye, but rather in the ring of maximum winds on the right side of the best track prepared for this cyclone. The radius of the maximum winds appears as less than 20 km in the forenoon of 9 June and the same becomes 25 to 30 km at 06 to 09 UTC, when the storm passes through Kandla. These inferences are just tentative, as they are based on scanty reports of very strong winds reported from a very poor network of observatories, which had no continuous wind-recording instruments except at Jamnagar.

According to the post-storms survey report carried out by the IMD, a very large number of windmills, which had tolerance potential of about 180 kmph, were completely damaged in the storm in the Dwarka Taluka. The Building Materials and Technology Promotion Council (BMTPC) under the Ministry of Urban Affairs and Employment, also got a Rapid Damage Assessment of Cyclone Affected Areas of Kutch and Saurashtra prepared by the Taru Research and Information Network, New Delhi, The observations of the IMD Survey Team agree with their assessments. Figure 3 shows that the strongest winds, about 200 kmph, are seen along and west of the track, particularly in the Dwarka Taluka of Jamnagar district. The gale force wind area is however quite extensive and covers a large area east of the track. Such an asymmetric distribution of winds was also seen by Raghavan (1990).

The casualities encountered in this cyclone are basically attributable to the storm surge of about 2 to 3 meters above the astronomical tide of about 3.2 meters in the Kandla port area. Dube *et al.* (1985) and Ghosh *et al.* (1983) have computed the storm surges encountered in west coast cyclones. A coarse-resolution storm surge model of IIT New Delhi for the west coast did not resolve the response of the Gulf of Kutch for this storm surge, and the computation made in respect of this cyclone yielded poor results. However, the model has been further refined, and the resolution has been increased to 5 km. According to the fine-resolution model of Dube *et al.* (personal communication) for this cyclone, the storm surge works out to 3.3 meters in the Kandla area on 9 June 1998, which is close to the estimates of the survey teams of both the IMD and BMTPC.

A devastating cyclone struck the Orissa coast of India near Paradip, on 29 October 1999. It was termed as a Super TC with an estimated wind speed of over 250 kmph that made landfall around 06 UTC on that day and paralyzed the entire state killing more than 9887 people. A huge surge of about 5 to 6 m above the astronomical tide was experienced along with very heavy rains, in which the cumulative total over the course of the storm period was 60 cm at some places! This storm surge penetrated the interior up to a distance of about 25 km from the coast. Plate 3 shows the impact of this cyclone. The Balasore area experienced the worst floods in the history of the last 100 years. About 35 villages in the Erasma block were completely washed out by the storm surge. The death toll in this region accounts for 80% of the total deaths of human beings in the state. The interesting aspect of this super cyclone is the fact that the peak storm surge was experienced very close to the point of landfall (within about 10–15 km), unlike the couple of previous super TCs in which it was observed at a distance of about 50 km. The radius of the maximum winds as inferred from the RMR was also of the order of 10 km and with this value of radius of maximum winds, the IIT model gives us a storm surge estimate of about 5.9 m. The probable maximum storm surge for areas close to and southwest of Paradip, as per Ghosh (1983), is little less than 4 m. The use of Das *et al.* (1974) nomograms gives us a value close to 6 m near Paradip. It is also interesting to see that the extent of the gale wind force was confined in the coastal belt between Gopalpur to the south and Chandbali to the north as seen in Plate 3. An extensive damage survey has been carried out with respect to this super TC by the Structural Engineering Research Center (SERC), Madras. The electric and telephone poles between Pure and Balasore on the east coast were damaged. Due to this, more than half of Orissa, 8000 telephones became dead. Electric poles were damaged and there was no power supply, even to the capital city, Bhubaneswar, for more than a week. Severe damage to buildings occurred in and around Paradip. The super TC caused catastrophic damage to well-engineered structures, industrial and port structures, microwave towers, communication and power transmission systems.

Other issues of Disaster Management

Warning System

The goal of any warning system is to maximize the number of people who take appropriate and timely action for the safety of their life and property. All warning systems start with the detection of the event and end with people getting out of harm's way. Such warning systems encompass three equally important elements namely: Detection and Warning, Communication and Response. We have discussed the detection process in detail above. Some relevant information on warnings, communications, and the response system follows.

The Four-stage Warning System

A two-stage warning system for tropical cyclones was in vogue in IMD, in which warnings were issued to state government officials in two stages: the first stage warning known as "Cyclone Alert" issued at least 48 hours in advance of the expected commencement of adverse weather over the coastal area and the second stage warning known as "Cyclone Warning" issued at least 24 hours in advance. These warning continue to be issued at 3-hourly intervals giving the latest position of the cyclone, its intensity (maximum sustained surface wind speed) and likely time and point of landfall, together with the storm surge height and type of damage expected.

In addition, with effect from the beginning of the cyclone season of 1999 (April–May), a special bulletin called the *Pre-cyclone Watch* has been introduced. This bulletin contains early warnings about the development of a cyclonic disturbance in the north Indian Ocean, its likely development into a tropical cyclone and the coastal belt likely to experience adverse weather is also indicated. This early warning bulletin is issued by the Director General of Meteorology and is addressed to the Cabinet Secretary and other Senior Officers of the Government of India, including the Chief Secretaries of the concerned maritime states.

Finally, to cover the devastating impact of the cyclones on the inland areas, a fourth stage of warning on the 'post-landfall scenario' has also been introduced. This commences about twelve hours before landfall and continues till such time as the cyclone-force winds are expected in the interior areas. At this stage, the district collectors of all interior districts besides the coastal areas likely to be affected are included in the bulletin. After the weakening of the cyclone into the depression stage, a final message on 'de-warning' is issued to all the concerned people.

Cyclone Bulletins and Warnings

The following bulletins and warnings are issued by the Cyclone Warning Centres for their respective areas of responsibility:

- Sea bulletins for ships plying in coastal waters;
- Broadcast of forecast/warning bulletins through INMARSAT, EGC Safety Met. Service, under the new Global Maritime Distress Safety System (GMDSS) Program of the World Meteorological Organization (WMO), for the ships in the Indian Ocean (north of the equator) have started w.e.f. 01-06-1996. This bulletin is issued at 0900 UTC every day. In addition, a second GMDSS bulletin is also being issued at 1800 UTC w.e.f. 01-10-1998:
- Bulletin for the Indian Navy;
- Port warnings;

- Fisheries warnings;
- Four-stage warnings for Central and State Government Officials;
- Bulletins for broadcast through AIR for the general public;
- Warning for registered recipients of the warnings;
- Bulletins for the Press;
- Warnings for Aviation;
- Bulletins for ships in the high seas through P & T's coastal radio stations.

During disturbed weather in the Bay of Bengal and the Arabian Sea, the ports likely to be affected are warned through post warning signals. The IMD also issues a "Fleet forecast" for the Indian Navy and Fisheries warnings for the use of fishermen during periods of disturbed weather. For the benefit of ship operators, two different types of bulletins are issued: coastal bulletins for the Indian coastal area covering up to 75 km from the coastline and sea area bulletins for the Arabian Sea and Bay of Bengal. Details of these warnings and bulletins are contained in Chapter 9 of the Cyclone Manual of the IMD.

Communication

This implies more than simply dissemination. Communication is complete only after the information is received and understood. An effective communications system is vital component of any effective warning system. Local officials and the public must first hear or see hazard warning information before they are in any position to take avoidance action.

The department makes use of 97 point-to-point teleprinter links connecting different field offices. These T/P lines are leased out from DOT/MTNL, whose responsibility includes their maintenance. In addition, a total of 69 centers have been provided with 85 telex connections. Switching computers have been provided at 5 Regional Collecting Centers for reducing handling delays. These computers are linked to the central Regional Telecom Hub (RTH) computer, New Delhi, through high-speed (2400 bps) links. The RTH computer system has been suitably upgraded. It is linked to major meteorological centers through 15 international circuits. This linkage helps in the exchange of global data, which is an essential component in the monitoring, analysis and forecasting of natural hazards. The RTH computer system can support 128 circuits, each upto a speed of 128 kbps, and can handle more than 1500 MBS of data per day. The system fully supports a state-of-the-art network technology and utilities like TCP/IP protocol, FTP, TELNET, Remote log-in, Mail facility, etc.

Radio Teletype (RTT) broadcast is another mode of data reception at field forecasting offices. In case of breakdown of the conventional T/P, it provides an effective means for the transmission of data and bulletins. About 27 field offices have since been provided with this facility. Due to international commitments under the WMO, India maintains RTT broadcasts to cater to regional requirements also. Radio Facsimile is another mean of communication for exchange of processed information. This broadcast can be heard within a distance of 6000 km; 28 field offices are equipped with this facility.

IMD also utilizes VSAT technology—most effective when conventional links break down. It operates at 1200 bits/sec and is comparatively cheaper than telex. This facility exists at 20 field offices. In addition, there are a number of HF/RT and VHF links. VHF links are effective for short-distance wireless communication or inter-city communication. A total of 65 telefaxes – an easy way of transporting processed information, documents and pictures using STD facility – have been provided. A meteorological data dissemination system has also been implemented.

The IMD has launched its own website (www.imd.ernet.in), where a large number of weather-related products including severe weather warnings are hosted for access by various agencies and public. Cyclone warnings are communicated to Government officials and other parties by high-priority telegrams, T/Ps, telex and telephones. The police wireless is also used as and when possible. The general public, the coastal residents and fishermen are warned through State Government officials, broadcast of warnings through AIR, and telecast programs in the national and regional hook-up. The cyclone warnings are issued to State Government officials in two stages.

Though the information on the pre-cyclone watch stage is faxed to the Chief Secretary, according to the "Four-Stage Warning Scheme", the collectors of coastal and few immediate interior districts along with the Chief Secretary of the concerned maritime State are warned from the cyclone alert stage onwards, whenever any coastal belt is expected to experience adverse weather (heavy rain/gales/tidal wave) in association with a cyclonic storm or a depression, likely to intensify into a cyclonic storm.

The "Cyclone Alert" is issued 48 hours in advance of the expected commencement of adverse weather over the coastal areas. Forecasts of the commencement of strong winds, and heavy precipitation along the coast in association with the arrival of a cyclone are issued at the alert stage. The landfall point is usually not identified at this stage. The "Cyclone Warning" is issued 24 hours in advance. The landfall point is forecast at this stage of the cyclone warning. In addition to the forecasts for heavy rains and strong winds, the storm surge forecast is also included at this stage. Since the elevation of the sea level is the biggest killer so far as the devastating attributes of a storm are concerned, information in this regard is most critical for taking follow-up action on evacuation of people from the low-lying areas likely to be impacted by the storm. A large coastal belt (roughly 500 km) is usually warned in the United States of America. It is creditable that a smaller strip was identified in Gujarat in the case of the 9 June 1998 cyclone.

During disturbed weather over the Bay of Bengal and the Arabian Sea, the ports likely to be affected are warned through "Storm Warning Signals". The department also issues "Fleet Forecasts" for the Indian Navy and coastal bulletins for the Indian coastal areas, covering upto 75 km from the coastline. Delhi Doordarshan Kendra introduced a new information service name "Teletext Service" for the benefit of viewers w.e.f. 14 November 1985. Meteorological information/forecasts of some major national and international stations are being provided for this program.

As a responsibility to the Regional Communities under the WMO/ESCAP panel on tropical cyclones, tropical cyclone advisories are issued by the Regional Specialized Meteorological Centre (RSMC) New Delhi, to the panel Member Countries during the tropical cyclones in the Bay of Bengal and the Arabian Sea.

In addition to the existing modes of dissemination of cyclone warnings to various State governments, P&T officials, port officials, etc through high-priority telegrams, telephones and telex by IMD, a new scheme, the Cyclone Warning Dissemination System (CWDS) using INSAT, was implemented on an experimental basis for coastal areas of south Andhra Pradesh and north Tamil Nadu in December 1985, by establishing 100 CWDS receivers. This scheme was made operational during the cyclone season of 1986–87. The state government officials of Andhra Pradesh found this scheme useful, especially in rural areas. This communication method is more reliable as it does not use terrestrial links which are disrupted during severe weather conditions. The cyclone warning message was to be originated from ACWC Chennai whenever a storm is observed for reception, by CWDS receivers located in the areas likely to be affected by the cyclone. After the successful demonstration of this scheme in the limited areas of Tamil Nadu and Andhra Pradesh, this scheme has been extended to other cyclone-prone coastal areas. So far, about 250 CWDS receivers have been installed in the cyclone-prone areas of the east and west coasts.

In times of need, the IMD makes use of the police wireless, coastal radio of DOT, CAD AFTN links, and railway microwave links. It also broadcasts two bulletins for safety at sea using the GMDSS system, operating through INMARSAT.

The WMO GTS provides the backbone of the telecommunications for the relay of warnings, forecasts, observational data and related information within the meteorological community and to some major external users and, in some instances, supports early warning for non-weather hazards. However, some of the other communication systems cited above are more appropriate for distributing warnings to the local population and external agencies, particularly when speed is essential. In general, the effective dissemination of warnings to the public and lower level administrators requires communication systems which have a very broad public reach, such as radio and television stations and community warning facilities. The dissemination of warnings through these external agents and facilities is, however, carefully coordinated to ensure timeliness and accuracy and, as noted elsewhere, experience confirms that there must be a single official issuing authority for warnings to minimize confusion.

Although tropical cyclone motion forecasts have improved over the years, the intensity forecasts have still a lot to be achieved. Using the frequently (hourly to half-hourly) available satellite imagery, a lot of experience has been gathered in the analysis of tropical storm intensity. Since the launch of INSAT, the intensity of cyclones has been well-captured. This has enabled cyclone experts to issue realistic storm surge forecasts. However, our knowledge of structural changes in tropical storms on account of land interactions is still in its infancy. The development of localized zones of strong winds, tornadoes and very heavy rains in some pockets continue to be elusive. There has been a vast improvement in forecasting techniques and cyclone warning services in recent years. This helped in considerable reduction of the loss of human lives and damage to properties caused by cyclones. The warning services provided by the department in recent years have been highly acclaimed by both the Government and the media.

Response

Through extensive studies of human response to disasters, it has been shown that warnings by themselves are not a stimulus–response reaction. Normally, people in a threatened area will first assess their personal risk. The additional information required before they take action depends on: (1) the content and clarity of the initial message, and (2) the credibility of the issuing organization. The potential for individuals to act are markedly increased if they are provided information that will: (1) define their risk, and (2) highlight what life-or property-saving action should be taken. The needful is done to ensure that the warnings generate the desired response at different levels. Though an individual is responsible for taking safety measures for his life and property, the state can not dissociate itself altogether from measures aimed at ameliorating the sufferings of its citizens. The directions of response at the State level (both central and state government level) are determined by the gravity of the situation. The administrative response, which may be driven by the political response, involves activating the administrative machinery, the success of whose ameliorative actions would depend upon the timely forecasting and efficient operation of the warning system. Institutional arrangements must exist between the Meteorological Organization and the crisis managers. Such an arrangement already exists between IMD and Disaster management Agencies both at the Center and the States. Crisis Management Groups have been constituted at the

Central and State levels, in which representatives from IMD provide the most critical information for suitable follow-up actions. There are three Area Cyclone Warning Centres (ACWCs) established at Kolkata, Chennai and Mumbai that cater not only to the requirements of cyclone warning in the respective States in which they are located, but also exercise technical control on the activities of the Cyclone Warning Centres (CWCs) established at Bhubaneshwar, Viskhapatnam and Ahmedabad. The cyclone warning work is coordinated and supervised by the office of the Deputy Director General for Weather Forecasting, Pune. A Cyclone Warning Division has been set up at the IMD HQ at New Delhi to advise the government at the apex levels on matters related to cyclones and also to the work of the Regional Specialized Meteorological Centre (RSMC) New Delhi, to provide cyclone advisories to all the Maritime States bordering the north Indian Ocean under the WMO/ESCAP panel. Crucial information from the concerned meteorological office flows down to the Control Rooms set up in this regard, both at the Center and the States, where Cyclone Distress Mitigation Committees (CDMCs) have also been established.

Meteorologists and hydrologists are not the sole members of the warning processes; rather they form a part of the larger hazard community made up of all organizations charged with the response to natural hazards. The information and technical assistance that the meteorologists can provide is absolutely critical to the warning process. Over the last 4 decades, research on community warning capability for disaster has revealed that effective public warnings are the product of the fully-managed team work of a combination of organizations. Someone has to take the lead to organize the community and develop a warning system. Meteorologists and hydrologists may not be able to complete the warning process by themselves, but they can take the lead in organizing hazard committees to ensure that effective systems are developed and maintained. In this connection, it must be mentioned that based on the input supplied by the IMD and the National Crisis Management Committee (NCMC), meetings were organized during the epochs of Tropical Cyclones in 1988 and 1999 to take stock of the situation for follow-up action. These coordination meetings worked very well and served the purpose for which they were conducted. Different organizations have different critical roles to play in the warning process. The meteorological office has a responsibility in the detection and prediction of natural hazards. Other organizations such as the Civil Defence or Emergency Management are responsible for issuing local warnings or alerts and still others like relief agencies may coordinate and monitor the public response to official warning. Each member of the multi-organizational warning network should pursue the overall warning goal. The communication of the critical information to the public is one component of the goal, the overall goal is to get the greatest number of people out of harm's way.

Apart from the many dos and don'ts listed by Siromony (2000) for people to meet the challenge posed by cyclones, it may be desirable to include here some of the actions most needed in the post-landfall scenario. There is a strong and imperative need for the district administration to be vigilant and to convene the district level committee to review the situation in the aftermath of the cyclone. This is intended not only to provide relief to the cyclone victims and to remove dead bodies and carcasses of the dead animals and to make arrangements for their disposal, but also to take measures to prevent epidemics and to ensure that the injured have been removed to hospitals and that an adequate number of mobile squads have been mobilized to clear the fallen trees. Immediate action is required to restore the power supply and ensure that there are no hanging wires or damaged electric poles lying on the ground. Damage reports are to be prepared immediately. Photographs of the affected areas, properties and people should be arranged, and contractors with adequate manpower and equipment should be pressed into service to restore normalcy at the earliest. The people in the affected area have to remain inside elevated shelters especially prepared for this purpose until informed by those in charge that they can return home. They should get themselves inoculated against diseases immediately at the nearest possible hospital and seek medical care for the sick and injured. They should keep away from the disaster area, unless called upon to assist in the process of disaster management. It is, however, the duty of the State and District administration to take adequate measures to protect the treasured possessions of the affected people who have been forced to vacate their dwellings.

Weakness in the Warning System

Apart from the problems arising on account of limitations, the major deficiencies that impede progress are:

- The lack of data to support warning and forecast services,
- Insufficient computer capabilities to implement more refined numerical models,
- The lack of communication capabilities within the hazard community and insufficient resources for warning coordination, public education, and advanced hydrometeorological training,
- Insufficient resources to support multi-agency, preparedness-planning efforts.

The IMD, with its limited resources, is making a great effort to improve the monitoring and forecasting of natural hazards. Some of the ongoing schemes aimed at improving our facilities are included in the following section. However, as already emphasized above, a well-planned coordinated effort in consultation, and possibly collaboration, with other concerned agencies is required.

Future Scenario

One could make a qualitative assessment of the social and economic benefits that could accrue from the Weather Warning Service. In documented cases, the cost/benefits ratios range from 1:55 to 1:127. The evidence suggests that every rupee spent in being prepared can save a hundred rupees or more that would otherwise be spent to make good after the event. Nowhere else are these very favorable cost/benefit ratios more obvious than in the cyclone warning service. Keeping this aspect in view, the IMD has always tried to update the technology required to enable the cyclone warning service to reach the state-of-the-art level. The IMD has ongoing projects for future development in the field of Space Meteorology, Telecommunication and Instrumentation. The ground segments for reception and processing of data from I METSAT & INSAT-3A satellite systems, likely to be launched in the first quarter of 2002, are being upgraded. We have planned to modernize the Meteorological Data Dissemination System by replacing the current Analogue System with a Digital System. A digital CWDS system is under implementation with one uplink earth station at RMC Chennai. A network of 100 CWDS receivers will be installed under this scheme in the coastal areas of Andhra Pradesh. There are also plans to make refinements in the Data Collection Platform (DCP) scheme. It is also proposed to install 4 more Doppler Radars in addition to the 6 Doppler Radars already approved. The IMD has plans to further modernize its telecommunication network to ensure that the reception of observations and dissemination of processed products is made without any delay. The existing message-switching systems at Mumbai, Kolkata and New Delhi will be shortly replaced by a state-of-the-art system, which will be capable of handling processed data and products transmitted through the newly-installed RTH computer system. There are also plans to expand the existing VSAT network substantially to include its provision at far-flung observatories in the country. The computing facility is also being suitably augmented to run high-resolution mesoscale models.

The Government of India has constituted a **High Powered Committee (HPC)** to review the existing norms of the Disaster Management Scheme and come up with several suitable suggestions to improve effective disaster management practices at the Center and the State levels. One of the ideas being floated is to set up a Special Task Force equipped with the latest technology and a nation-wide computerized database. The community has to be increasingly involved in the process of disaster management and, to create greater awareness in the community, voluntary organizations such as NGOs, have to step in.

Gaps in the Existing Systems and Suggestions for Improvement

The main problem areas are listed below:

- There is an inadequate observational network on the high seas and along the coastline. Thus, there is a strong requirement for strengthening the observational network, both the surface and upper air, including high wind-speed recording instruments. There is also an imperative need for augmentation of ocean observing systems (Ocean Data Buoys) in the high seas (north Indian Ocean) for the detection and monitoring of tropical cyclone formation and movement.
- The currently operational INSAT satellites have proved very useful for tropical cyclone analysis and prediction. For effecting further improvements, satellite systems with multi-channel image capabilities of higher resolution would improve the current techniques of analysis and forecasting. The INSAT satellite systems should have an automatic image navigation and registration, so that better quality products could be derived without delay. Microwave imagers have also been useful and have proved their ability. High-frequency sensors (37 and 85 GHZ) need to be flown on board Indian polar-orbiting satellites and their data should be analyzed on a real-time basis. They should also have sounding capabilities at higher resolution (20 km).
- The deployment and networking of an adequate number of Doppler Radars would improve the analysis and prediction of cyclones. To start with, all the conventional radars should be replaced with Doppler Radars.
- Improving the design of tide-gauges is essential to capture storm surges and augmentation of their network along the east and west coasts of India.
- The cyclone forecasting in IMD is done by state-of-the-art techniques using conventional synoptic methods, supported by computer-generated numerical guidance information, available from dynamical weather prediction models. IMD runs its own numerical weather prediction (NWP) limited area model for cyclone track prediction, for which special techniques have been indigenously developed in IMD. Forecasts available from advanced NWP centers from other parts of the world like the European Centre for Medium Range Weather Forecasting, NCEP Washington, etc. are also looked into. Though our forecasting capability has tremendously improved in recent times with the availability of modern observational and forecasting tools, the accuracy of the forecasts of movement of cyclones and their landfall information with regard to location and time, which is very crucial from the disaster management point of view, still suffers from some inherent limitations. As a result, a large belt of the coast extending up to a few hundred kilometers has to be necessarily covered by the cyclone warning

bulletins issued by IMD. It is imperative to modernize the computerized weather forecasting system of IMD, by introducing very fine-resolution numerical models for tropical cyclone predictions.

- To facilitate the crisis managers and public administrators, relatively longer range (> 48h) forecasts of tropical cyclone tracks and all other attributes of tropical cyclone may be required. This may require ocean–atmosphere coupled models, with suitable parameterization schemes, to comprehend the building-up of intense precipitation strong winds and storm surges well in advance. This would essentially mean upgrading the computing facility in IMD with high-power computer systems, and this would also require human-resource development.

Management System

A good network of motorable roads should be constructed in all vulnerable coastal areas. This not only facilitates quick evacuation in the time of need, but also the supply of relief to the needy in the aftermath of the cyclone (for example, the East Coast Highway).

We should construct a high-level coastal road with its top about one meter above the highest surge level which should, of course, have adequate drainage openings to permit the flow of normal rain/floodwaters across. Such a road could serve as the first line of defense against the surge waters, permitting only limited entry of water through the openings. Or, alternatively we should build a cyclone super-highway at a distance of about 100 km from the coast and concentrate all economic activity in towns built on this highway.

Our national highway in the east almost hugs the east coast and becomes unserviceable during the cyclone landfall. The Cyclone super-highway will become a life-line during and immediately after cyclone hits.

Recent experience with the super cyclone that struck Orissa state near Paradip on 29 October, 1999, shows that virtually nothing happened to *pucca* buildings. Therefore, schools and hospitals may be built on the super-highway and these may be used as cyclone shelters during the cyclone.

Training and educating the people in the cyclone-prone areas is a very important component of disaster management machinery, so as to prepare them to mitigate the disaster themselves. This will create an awareness about the disaster warning systems and the vital importance of the cyclone warnings. They should also be taught to ensure proper utilization of the cyclone shelters. Community preparedness and the participation by local people is required for the successful implementation of any plan of action for cyclone disaster dealing. Alerting the people on the "Dos" and "Don'ts" through education, the media, press and TV/radio is also required.

In one of the recent surveys, it came to our notice that some fisherman go fishing at the time of the cyclone with the hope of getting a bigger catch of fish, which is their livelihood. In order to prevent such activities, the proposal of supplying free, dry rations to those fishermen likely to bear the brunt of the cyclone may be considered.

A reliable communication system based on state-of-the-art-technology is an essential requirement for an effective warning system. For the dissemination of the warning messages, communication systems which have very broad public reach such as radio, television, and community warning (cyclone warning dissemination system—CWDS) should be used. The use of local warning systems such as sirens, loudspeakers, door-to-door visits, which are community-based approaches, should be maximized. For the dissemination of warnings to the fishermen at sea, direct satellite broadcast (such as Worldspace) can be employed.

Venkatachary (1998) emphasized the use of space technology for disaster management. There is also a need to have connectivity between the different departments (IMD, CWC, DOT, DOS, S&T) for the exchange of data to ensure timely and effective service. The type of data and the quantum of data that need to be exchanged between the different departments have to be worked out and finalized.

The role of NGOs will be in organizing relief operations and rehabilitation. They can create opportunities for proper socio-economic development and help in capacity building of the people. Efforts may also be made to promote self-reliance of the peoples' organizations and to network them for collective action on issues of concern. The training and orientation initiative of various NGOs, linked with their development interventions, have enabled people to appreciate the significance of disaster preparedness as a conscious and planned effort to adequately respond to emergencies.

We also need to develop robust systems with minimum maintenance requirements, easy to operate, with software that is user-friendly and communicative to different levels of users. The warnings have to be understood by the affected persons, free of technological jargon, easy to act upon and have continuity in terms of content and meaning. Also, there should be a standby communication system in the event of failure of the main electric supply.

ACKNOWLEDGEMENTS

The author is extremely grateful to the Director General of Meteorology for providing him an opportunity to participate in the one day Seminar organized by INSA on Natural Hazards held at Vighyan Bhawan, New Delhi. He would also like to thank organizers for bringing out the proceedings of this seminar.

REFERENCES

Anderson, E and Hollingsworth, A (1988) Typhoon bogus observations in the ECMWF data assimilation system, *Tech. Mem.*, **148**, ECMWF, September, 1988.

Arkin, P A (1983) A diagnostic precipitation index from IR satellite imagery, *Trop. Ocean-Atoms. Newslett.*, **17**: 5–7.

Bhatia, R C, Manchanda, T C, Pradhan, D and Datta, R K (1999) Recent improvements in the radar network of IMD and future plans of upgradation of cyclone detection radars, *Meteorology beyond 2000: Proc. of TROPMET-99*: 165–169.

Chan, J C L and W M Gray (1982) Tropical cyclone movement and surrounding flow relationships, *Mon. Wea. Rev.*, **110**: 1354–1374.

Das, P K (1972) Prediction model for the storm surges in the Bay of Bengal, *Nature*, **239**: 211–213.

Das, P K, M C Sinha and V Balasubramanya (1974) Storm surges in the Bay of Bengal, *Quart. R. Met. Soc.*, **100**(425): 437–449.

Davidson, N E and K Puri (1992) Tropical prediction using dynamical nudging, satellite-defined convective heat sources, and a cyclone bogus, *Mon. Wea. Rev.*, **120**: 2501–2522.

Dube, S K, Sinha, P C, Rao, A D and Rao, G S (1985) Numerical modeling of storm surges in the Arabian Sea, *Appl. Math Modelling*, **9**: 289–294.

Dube, S K, Rao, A D Sinha, P C Murty, T S and Babulayan (1997) *Mausam*, **48**: 283–304.

Dvorak, V F (1975) Tropical Cyclone Intensity Analysis and Forecasting from satellite data, *Mon. Wea. Rev.*, **103**: 420–430.

Dvorak, V F (1984) Tropical Cyclone Intensity Analysis using satellite data, *NOAA, Tech. Rep., NESDIS*, **11**: 44.

Dastoor, A and T N Krishnamurti (1991) The landfall and structure of a tropical cyclone. The sensitivity of model predictions to soil moisture parameterizations, *Boundary Layer Meteorology*, **55**: 345–80.

Elsberry, R L (1998) Track forecast guidance improvement for early warnings of tropical cyclones, Paper presented at the International IDNDR-Conference on Early Warning Systems for the Reduction of Natural Disasters, Potsdam, Germany.

George, G E and W M Gray (1976) Tropical cyclone motion and surrounding parameter relationships, *J. Appl. Meteor.*, **15**: 1252–1264.

Ghosh, S K (1977) Prediction of storm surges in the east coast of India, IJMH&G, **28**(2): 157–168.

Ghosh, S K (1983) Probable Maximum storm surge on the coasts of India and Bangladesh, *Aspects of Mechanics*, pp. 74–92.

Ghosh, S K, Diwan, B N and Singh, V C (1983) Numerical simulation of storm surge envelopes associated with the recent severe cyclonic storms impinging on the east and west coast of India, *Mausam*, **34**: 399–404.

Gupta, G R, Mishra, D K and yadav, B R (1977) The Porbandar Cyclone of October 1915, *IJMHG*, **28**(2): 177–188.

Gupta, Akhilesh and Romesh, K J (1999) Prediction of tropical cyclogenesis over north Indian Ocean by Global Model, *Meteorology Beyond 2000, Proc. of TROPMET-99*: 215–225.

Holland, G J (1980). An analytical model of the wind and pressure profiles in hurricanes, *Mon. Wea. Rev.*, **108**: 1212–1218.

Holland, G J, McGeer, T and H Youngren (1992) Autonomous aerosondes for economical atmospheric soundings anywhere on the globe, *Bull. Amer. Meteor. Soc.*, **73**: 1987–1998.

Holland, G J (1984a) Tropical cyclone motion: A comparison of theory and observation, *J. Atoms. Sci.*, **41**: 68–75.

Iwasaki, T, H Nakano and M Sugi (1987) The performance of typhoon track prediction model with cumulus parameterization, *J. Met. Soc. Japan*, **65**: 55–570.

Joshi, P C, B M Rao and C M Kishtwal (1999) Prospects of meteorological satellites applications: India, *Meteorology Beyond 2000, Proc. of TROPMET-99*: 154–164.

Kalsi, S R and R K Jain (1989) On some Aspects of Marginal Cyclones, *Mausam*, **40**: 47–50.

Kalsi, S R (1993) INSAT image analysis of Tropical Cyclone development in the Bay of Bengal, in *Advance in Tropical Meteorology* (Eds. Keshvamurty, R N and Joshi, P C), Tata McGraw Hill, pp. 289–300.

Kalsi, S R, Rao, A V R K, Misra, D K, Jain, R K, and Rao, V R (1996) Structural variability of tropical disturbances in the Bay of Bengal, in *Advances in Tropical Meteorology* (Ed. by R K Datta), Concept Publishing Co., New Delhi.

Kalsi, S R and Gupa Manu (1998) Success and failure of early warning systems—a case study of Gujarat cyclone of 9 June 1998. Paper presented in the International IDNDR-conference on Early Warning Systems for the Reduction of Natural Disasters, Potsdam, Germany.

Kalsi, S R (1999) Some aspects of the Gujarat cyclone of June 1998, *Meteorology beyond 2000, Proc. of Tropmet-99*: 302–308.

Kalsi S R (2001) Role of IMD and Scientific Methods of Prediction of Tropical Cyclones and Advance Warning Systems, Paper presented in the National Seminar on Habitat Safety against Earthquakes & Cyclones, organized by HUDCO, BMPTC and MOUD, Govt. of India, from 2–3 May 2001, at Vigyan Bhawan, New Delhi.

Kelkar, R R (1997) Satellite-based Monitoring and Prediction of Tropical Cyclone Intensity and Movement, *Mousam*, **48**,2: 157–168.

Krishnamurty, T N, Arun Kuamr, Yap, K S Ashu, P Dastoor, Noel Davidson and Jian Sheng (1990) Performance of a high-resolution mesoscale tropical prediction model, *Advances in Geophysics*, Academic Press INC, **32**: 133–286.

McBride, J L and Gray, W M (1980) Mass divergence in tropical weather systems, Part I: diurnal variation. *Quart. J. Roy. Meteor. Soc.*, **106**: 501–516.

Mishra, D K and Gupta, G R (1976) Estimation of maximum wind speeds in tropical cyclones occurring in Indian Seas, *IJMH&G,* **27**(3): 285–290.

Mohanty, U C (1994) Tropical Cyclones in the Bay of Bengal and deterministic methods for prediction of their trajectories, Sadhana, *Acad. Proc. In Engg. Sci.,* **19**(4): 567–582.

Mohanty, U C and Gupta, Akhilesh (1997) Deterministic methods for prediction of Tropical cyclone tracks, *Mausam,* **48**(2): 257–272.

Mukherjee, A K and Subramanian, S K (1977) Some features of the Porbandar Cyclone of October 1975, *Mausam,* **28**(4): 439–446.

Prasad, K Sanjib Sen, Joardar, D and Rama Rao, Y V (1995) Operational Numerical Weather Prediction system for Short-range Forecasting in India Meteorological Department, *Proc. of the 5th Regional Workshop on Asian/ African Monsoon Emphasizing Training Aspects,* New Delhi, TMRP Rep. No. 52: 63–75.

Prasad, K, Kalsi, S R, Rama Rao, Y V, Ray Bhowmik, S K and Joardar, D (2000) Evaluation of Operational Cyclone track forecast by the IMD limited area NWP Model, *Mausam,* **51**(1): 91–95.

Raghvan, S (1990) Structure of tropical cyclones in the Bay of Bengal, *Mausam,* **41**(2): 325–328.

Raghvan, S (1997) Radar Observations of Tropical Cyclones over the Indian Seas, *Mausam,* **48**(2): 169–188.

Rai Bhowmik, S K and Prasad, K (2000) Some characteristics of Limited-area Model Precipitation forecast of Indian monsoons and evaluation of associated flow features, *Meteorl. Atmos. Phys.,* pp. 1–14.

Raman, S (1965) The severe cyclonic storm of 11–12 June 1964 in the Arabian Sea and some of its surface characteristics, *IJMH&G,* **16**(4): 623–630.

Ramathan, Y and Bansal, R K (1977) On the application of a primitive equation barotropic model for the prediction of storm tracks in the Indian region, *Indian J. Met. Hydrol. Geophys.,* **28**: 169–176.

Sarkar, A, Mohanty, U C Potty, K V J and Mandal, M (1999) Numerical simulation of tropical cyclones using a high-resolution regional model, *Proc. of INTROMET-97 on Asian Monsoon and Pollution over the Monsoon Environment,* Vayu Mandal, pp. 13–16.

Siromony, Michael, P V (2000) *Source Book on District Disaster Management,* Ministry of Agriculture Department of Agriculture and Cooperation, Govt. of India.

Sikka, D R (1975) Forecasting the movement of tropical cyclones in the Indian sea by the non-divergent barotropic model. *Indian J. Met. Hydrol. & Geophys.,* **26**: 323–25.

Veldin, C S, Olson, W S and Roth, B A (1989) Tropical Cyclone center-fixing using SSM/I data, Preprints, *4th Conf. Sat. Meteor. Oceanogr., Amer. Met. Soc., Boston, MA 02,* **108**: J36-J39.

Venkatachary, K V (1998) Use of space technology for disaster management, ET SAMTA Workshop, ISRO, SAC, Ahmedabad.

Weatherford, C L and Gray, W M (1988) Typhoon structure as revealed by aircraft reconnaissance, Part II, Structural Variability, *Mon. Wea. Rev.*, **116**: 1044–1056.

Wexler, H (1947) Structure of hurricanes as determined by radar, *Ann. N.Y. Acad. Sci.*, **48**: 821–844.

Willoughby, H E, Close, J A and Shoreibah (1982) Concentric eyewalls, secondary wind maxima and the evolution of the hurricane vortex, *J. App Sci.*, **39**: 395–411.

Additional references on Cyclones and Cyclone related disaster management

Annual Review of Cyclones in the North Indian Ocean, Ed. R R Kelkar, Printed by WMO, available for the years 1997, 1998 and 1999.

Anti-disaster Plan for Tamil Nadu, September 1978.

Cyclone Manual, Chapters 1 to 12 (IMD internal publication).

Final Report of the Cyclone Distress Mitigation Committee, Andhra Pradesh, Hyderabad, 1971.

Report of the Cyclone Distress Mitigation Committee, Govt. of Orissa, Bhubaneswar, 1971.

Report of the Cyclone Distress Mitigation Committee, Govt. of West Bengal, Calcutta, 1974.

Report of the Cyclone Review Committee (Vol. **I**) for implementation of recommendations by Central Government Ministries, 1984.

Recommendations of Cyclone Review Committee (Vol. **II**) for implementation by Coastal State Governments.

Recommendations of Cyclone Review Committee (Vol. **III**) for implementation mainly by IMD.

Mausam, A Quarterly Journal of Meteorology, Hydrology & Geophysics, Vol. **48**(2), Special Issue on Tropical Cyclones. Ed Dr P K Das, 1997.

Mausam, A Quarterly Journal of Meteorology, Hydrology & Geophysics, Vol. **48**(4), Special issue on Ocean Response to Tropical Cyclones. Ed Dr Dube, S K, Octorber, 1997.

Natural resources management-A new perspective-Satellite remote sensing in disaster management. P P Nageshwar Rao and V Jayaraman, National resource management system, Department of Space, Bangalore.

Smith, D K (1989) Natural Disaster Reduction: How Meteorological and Hydrological Services Can Help, WMO-No. **722**, WMO, Geneva, p. 43.

Tracks of storms and Depressions in the Bay of Bengal and the Arabian Sea (1877–1970) India Meteorological Department, 1979.

Tracks of storms and Depressions in the Bay of Bengal and the Arabian Sea (1971–1996) Indian Meteorological Department, 1996.

United Nations (1976) *Disaster Prevention and Mitigation*, Vol. **2**, Hydrological Aspects, Office of United Nations, New York.

United Nations (1978) *Disaster Prevention and Mitigation*, Vol. **4**, Meteorological Aspects, Office of United Nations, New York.

United Nations (1978) *Disaster Prevention and Mitigation*, Vol. **10**, Public Information Aspects, Office of United Nations, New York.

Vulnerability Atlas of India: Earthquakes, Wind storms and Flood Hazard Maps and Damage Risk to Housing, Ministry of Urban Affairs and Employment, Govt. of India, 1997.

WMO (1990) *The Role of the World Meteorological Organization in the International Decade For Natural Disaster Reduction*, WMO-No. **745**, p. 37.

WMO (1991) *Tropical Cyclones and their forecasting and Warning Systems in the North Indian Ocean*, Ed G S Mondal, Report No. TCP-**28**.

WMO (1993) *Global Guide to Tropical Cyclone Forecasting*, Ed G J Holland, WMO TCP Report No. TCP-**31**.

WMO (1995) *Global Perspective of Tropical Cyclone*, Rep. No. TCP-**38**.

The impact of Tropical Cyclones on the coastal regions of SAARC Countries and their influence in the Region, SMRC-NO. **1**, SAARC Meteorological Research Centre (SMRC), 1998.

A New Seismic Hazard Map for the Indian Plate Region under the Global Seismic Hazard Assessment Programme

M Ravi Kumar and S C Bhatia*

ABSTRACT

A new seismic hazard map for the Indian plate region, comprising the Himalaya, North-east India, the Indian Shield, South China, Nepal, Burma and the Andaman Islands, was prepared under the Global Seismic Hazard Assessment Program (GSHAP). A working catalogue of the main shocks was obtained by merging the local catalogues from different countries, with the global catalogue of the NOAA. Eighty-six potential seismic source zones were delineated, based on the major tectonic features and seismicity trends. Using the probabilistic hazard assessment approach, the Peak Ground Accelerations (PGA) were computed for 10% probability of exceedence in 50 years, at locations defined by a grid of 0.5° × 0.5°. The PGA values over the grid points were contoured to obtain a seismic hazard map. The map reveals that the zones of highest risk are the Burmese Arc, North-eastern India and the Hindukush regions, with PGA values of the order of 0.35–0.4g. Also, a majority of the North Indian plate boundary region and the Tibetan Plateau region have a hazard level of the order of 0.25g. In the Indian Shield region, it is of the order of 0.05–0.1g, whereas some locales like Koyna depict a hazard level of about 0.20g.

* National Geophysical Research Institute, Uppal Road, Hyderabad, India

Introduction

Seismic Hazard Assessment

Seismic hazard, in the context of engineering design, is generally defined as the predicted level of ground acceleration, which would be exceeded with 10% probability at the site under consideration, in the next 50 years, due to the occurrence of an earthquake anywhere in the region. A lot of complex scientific perception and analytical modeling is involved in seismic hazard estimation. A computational scheme essentially involves the delineation of seismic source zones and their characterization, the selection of an appropriate ground motion, attenuation relation, and choosing a predictive model of seismic hazard.

Since earthquake catalogues constitute the first essential input for the delineation of seismic source zones and their characterization, the preparation of a unified working catalogue for a region under consideration is an important task. Globally speaking, the data from historic to the recent past can be broadly be divided in to three temporal categories:

1. from 1964 till the recent past, for which modern instrumentation-based data are available
2. from 1900 to 1963, the era of early instrumental data, and
3. pre-1900, consisting of pre-instrumental data, based primarily on historical and macro-seismic information.

The next key component of seismic hazard assessment is the creation of seismic source models, which demands translating seismotectonic information into a spatial approximation of earthquake localization and temporal recurrence. The seismotectonic maps need to be critically studied for defining arial seismic source zones and active faults. An earthquake recurrence model is then fitted to these source zones for defining the parameters that characterize the seismicity of the source region. These go as inputs to the algorithm for the computation of seismic hazard, viz., Minimum and Maximum magnitudes, and the parameters 'a' and 'b' in the earthquake frequency magnitude relationship: $\log N = a - bM$. The third main element required for seismic hazard assessment is the designation of strong ground motion (ground acceleration) estimation equations, specifying the ground acceleration as a function of earthquake magnitude and hypocentral distance. These equations have been developed for only a few regions of the world. Obtaining realistic estimates of strong ground motions in all regions is a major challenge that must be met.

Global Seismic Hazard Assessment Program (GSHAP)

Although the steps involved in hazard assessment are specific to the region under consideration, some standardization of the approaches is very essential so that the estimates can be compared worldwide and this will also to ensure consistency across the regional boundaries. GSHAP is a program, which has been co-ordinated internationally, and implemented at the regional and local levels, through a number of centers. It has adopted the probabilistic hazard assessment approach of McGuire to estimate the Peak Ground Accelerations (PGA), using FRISK88M software. Each center has been responsible for a defined geographical territory, for the preparation of a unified/ homogeneous earthquake catalogue, compilation of seismo-tectonic information and earthquake-source delineation, strong seismic ground motion studies, and the computation of predictive seismic hazard. The National Geophysical Research Institute (NGRI), Hyderabad, India, was identified as one such center, responsible for estimating the seismic hazard for the Indian region.

Seismic hazard assessment of the Indian region

The issue of seismic hazard in India has been addressed by the scientists as early as 1956, when a 3-zone (Severe, Moderate, Minor hazard) Seismic Zoning map of India was brought out (Tandon, 1956). This map was based on a broad concept of earthquake distribution and geotectonics. The severe hazard zones were roughly confined to the plate boundary regions, i.e., the Himalayan Frontal Arc in the north, the Chaman fault region in the North-west and the Indo-Burma border region in the North-east. While the minor hazard zone was confined to the Indian Shield region in the south, the moderate hazard zone was confined to the transitional zone in between the two. Since then, many versions of the seismic zoning map of India have been brought out. The Bureau of Indian Standards, which is the official agency for publishing seismic hazard maps and codes in India, has prepared a six-zone map in 1962, a seven-zone map in 1966, and a five-zone map in 1970/1984, which is currently valid. The present five-zone map is under review. Khattri *et al.* (1984), adapting a probabilistic hazard computational approach, published a map of seismic hazard in units of 'g', for 10% probability of exceedence in 50 years.

The present study was initiated during an international workshop of the GSHAP, held at NGRI in February–March 1996. The study region comprised the

Test Area #8 of GSHAP, covering parts of India, China, and Nepal, bounded by 20°N–40°N and 85°E–105°E. Towards the preparation of an earthquake catalogue for source zonation, the NOAA catalogue and several local catalogues were considered. A working catalogue of main shocks was prepared by removing duplicates, aftershocks, and earthquakes without any magnitude. To start with, the test area was divided into 16 source zones on the basis of the seismicity patterns emerging from a plot of epicenters in the region, in conjunction with the tectonic information. Later, further exercises were done with 30 source zones and 56 source zones. After a critical examination of the seismicity and tectonic constraints along with the computed PGA values, it was felt that the 30-source zone model is best suited for the region (Bhatia *et al.*, 1997).

Subsequently, the study was extended to a larger region bounded by 0-40°N and 65-100°E, which includes the entire Himalayan belt, North-east India, the Indian Shield, South China, Nepal, Burma and the Andaman Islands.

Source Zones and their Seismotectonics

To delineate the source zones based on the seismicity and tectonic information, a compilation of the tectonic features of the Indian region was made (Figure 1), based on a generalized tectonic map of India (Khattri *et al.*, 1984), a tectonic map of the Himalayan Arc (Khattri, 1987), a tectonic map of India published by the Oil and Natural Gas Commission (ONGC), a sketch map of the major tectonic features of South-east Asia (Leloup *et al.* 1995), a map of the Tibetan region showing fault plane solutions of moderate earthquakes and active faults (Molnar, 1992) and some unpublished material. The seismicity map of the study region for earthquakes of magnitude 4 and above is shown in Plate 4. Plate 5 shows the earthquakes of magnitude 6 and above, to highlight the regions that have experienced major earthquakes in the past. As can be seen, the seismicity of the Indian region is intense along its plate margins and is rather diffused in other regions, except for some concentrations in regions like Koyna. The tectonic and seismicity patterns of the Indian plate boundary regions have been studied in detail, using a large number of focal mechanism solutions of the Harvard CMT data (Kumar and Rao, 1995; Kumar *et al.*, 1996; Kumar *et al.*, 1998). Based on the seismicity and tectonic trends, 86 potential source zones were delineated (Plates 4 & 5), which are described below.

The Indian Plate Boundary Regions

The boundaries of the Indian plate are characterized by a continental collision segment along the Himalaya in the north, a complex

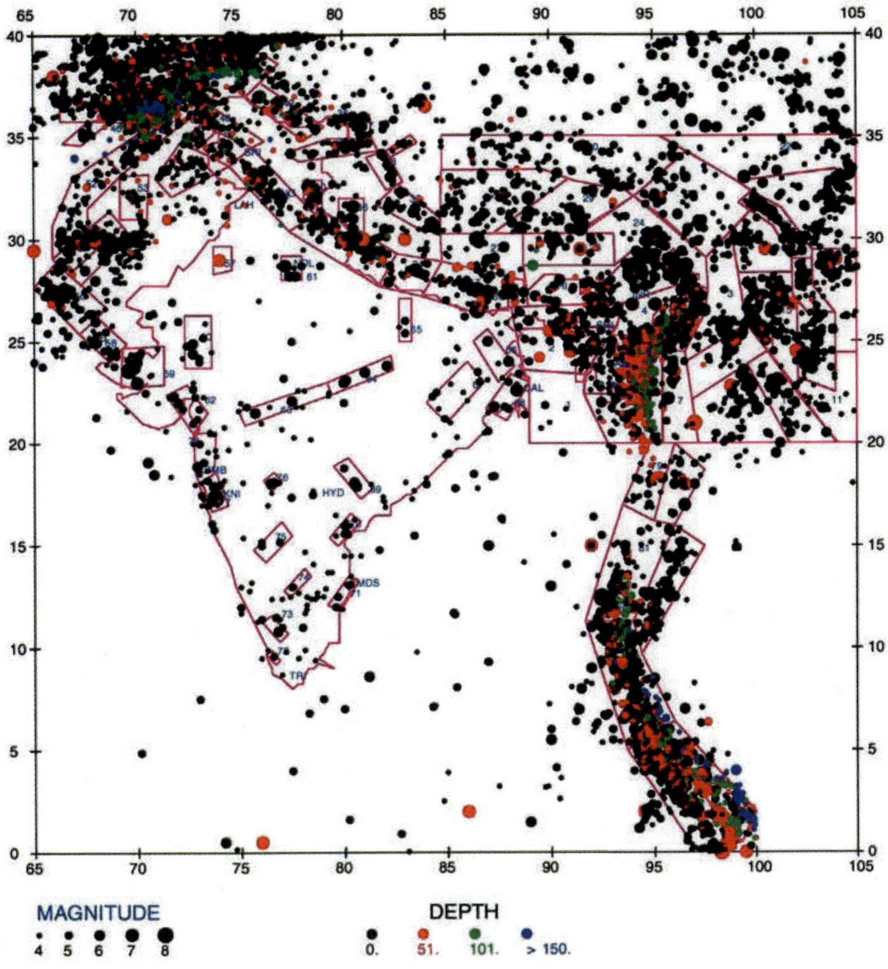

Plate 4 Seismicity and Source Zones Map of India and Adjoining Regions. The dots are the epicenters of earthquakes of Mag. 4 and above. Blocks numbered 1 through 86 are the source zones. Some cities are shown for orientation purposes: BMB (Bombay), BUJ (Bhuj), CAL (Calcutta), HYD (Hyderabad), JBL (Jabalpur), KAT (Kathmandu), KBL (Kabul), KOY (Koyna), LAH (Lahore), LAT (Latur), MDS (Madras), NDL (New Delhi), PBL: Port Blair, QUE (Quetta), SHL (Shillong), SRI (SriNagar), VSP (Vishakhapatnam).

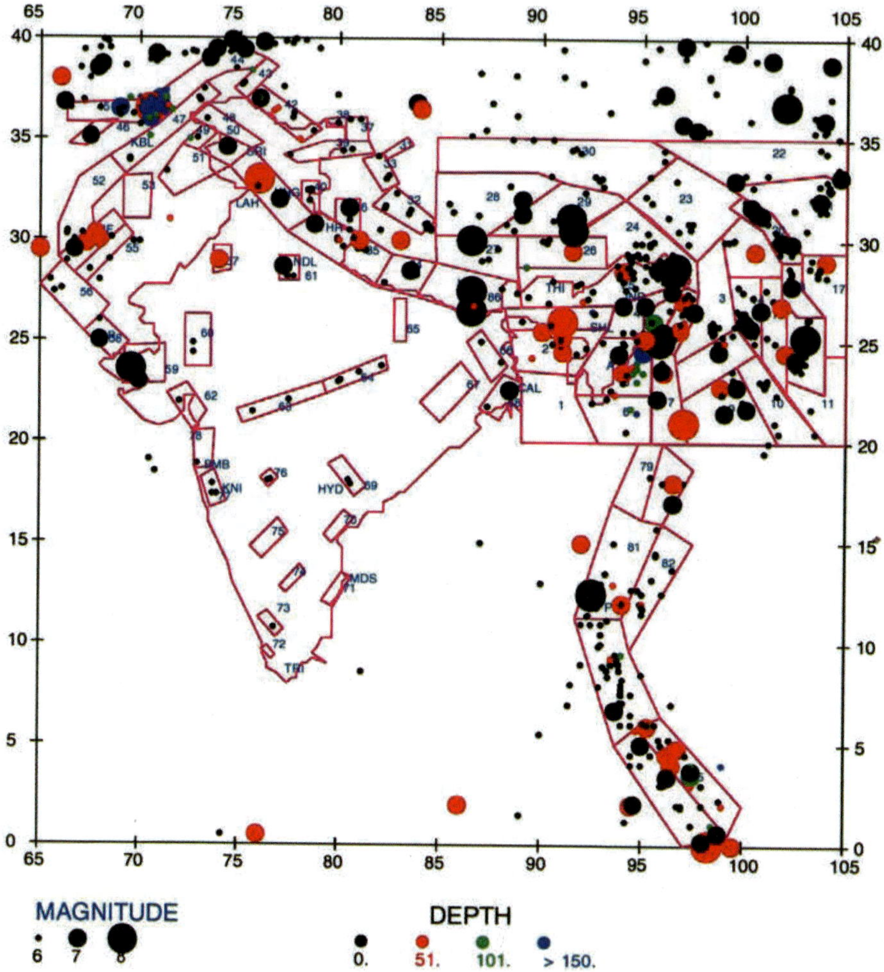

Plate 5 Map Showing Locations of Earthquakes of Magnitude 6 and above in India and Adjoining Regions. Three-letter locations refer to cities, as listed in the caption of Plate 4.

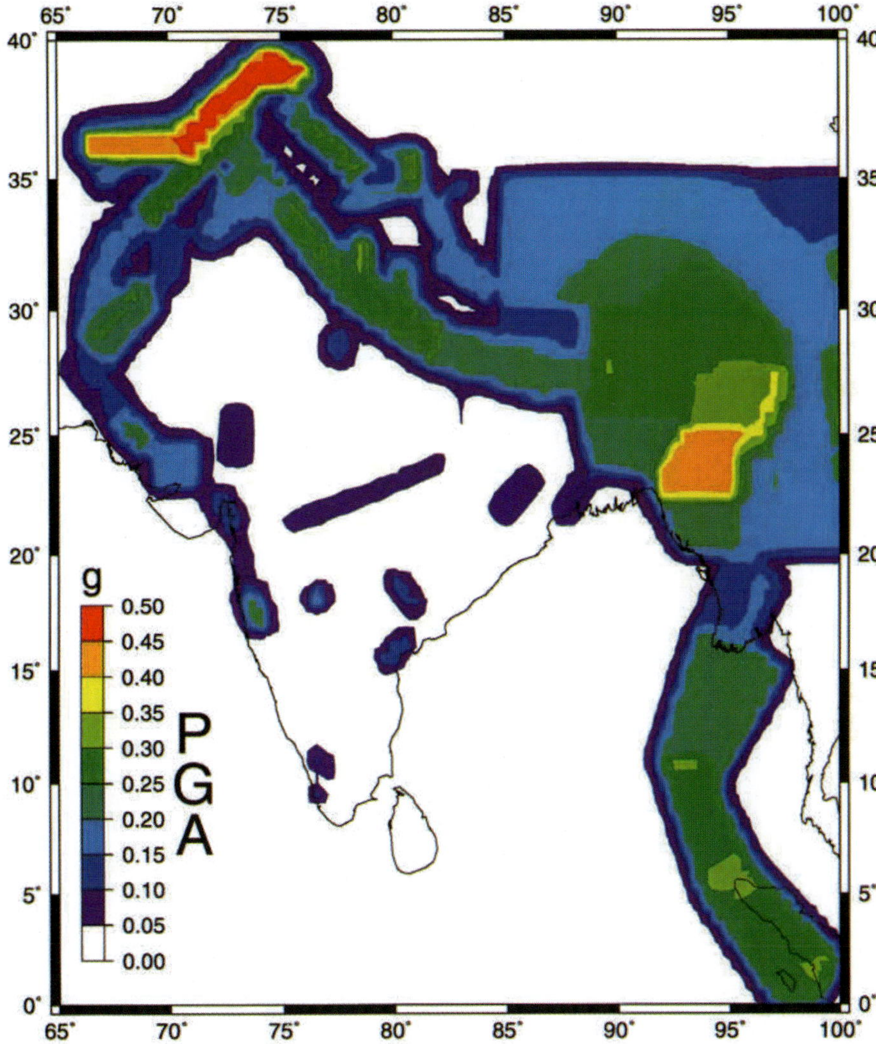

Plate 6 Seismic Hazard Map of India and Adjoining Regions for 10% Probability of Excedence in 50 years. Contour interval 0.05g.

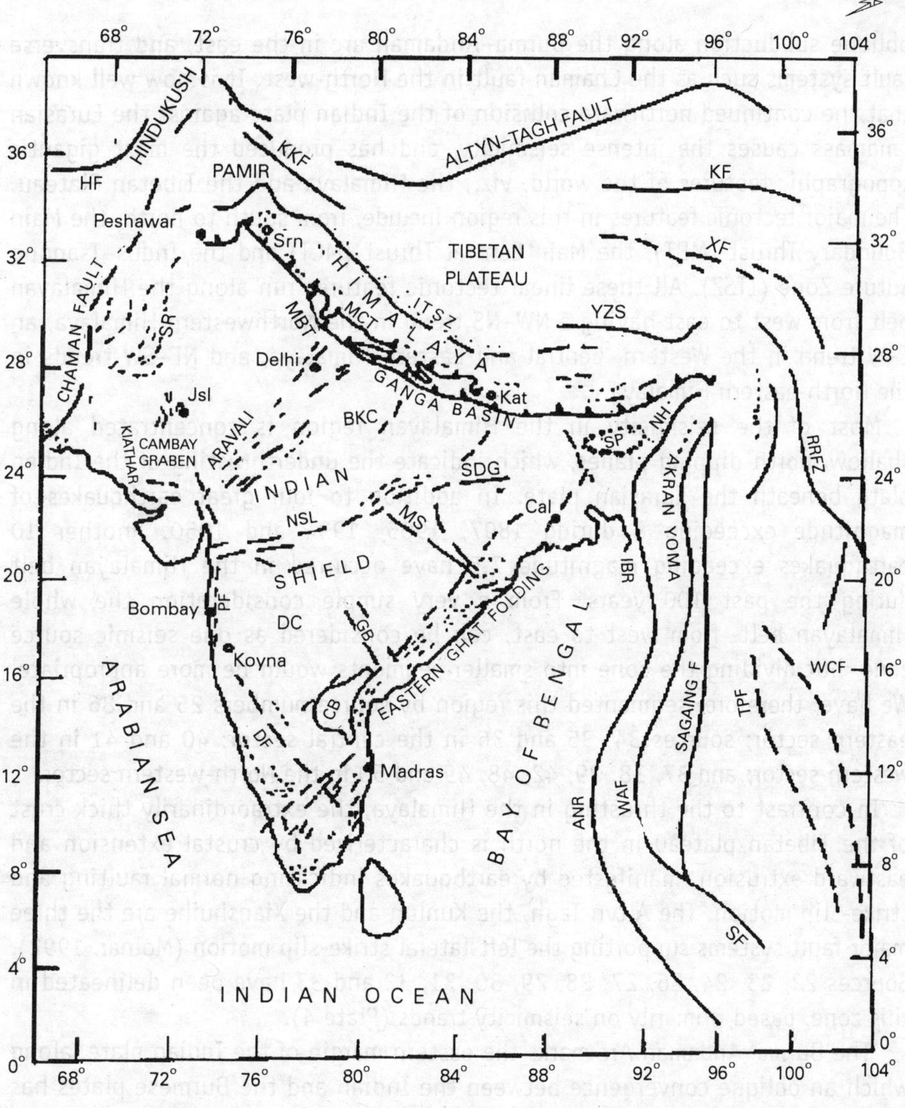

Figure 1 *Generalized Tectonic Map of India and adjoining Regions. Abreviated Tectonic features are ANR: Adaman–Nicobar Ridge, CB:Cuddapah Basin, DF:Dharwar Fold, GG:Godavari Graben, HF: Herat Fault, IBR: Indo-Burma Ranges, ITSZ: Indus Suture Zone, KF: Kunlum Fault, KKF: Karakoram Fault, MBT: Main Boundary Fault, MCT: Main Central Thrust, MG: Mahanadi Graben, SDG: Satpura–Damodar Graben, NH: Naga Hills, NSL: Narmada–Son Lineament, PF: Panvel Flexure, RRFZ: Red River Falt Zone, SF: Sagaing Fault, SR: Sulaiman Range, SP: Shillong Plateau, WAF: West Adaman Fault, WCF: Wang Chao Fault, XF: Xian Shui He Fault, YZS: Yarlung–Zangpo Suture. Cities shown are Cal: Calcutta, Jsl: Jaisalmair, Kat: Kathmandu, Srn: Srinagar.*

oblique subduction along the Burma–Andaman arc in the east, and transverse fault systems such as the Chaman fault in the North-west. It is now well known that the continued northward collision of the Indian plate against the Eurasian landmass causes the intense seismicity, and has produced the most gigantic topographic features of the world, viz., the Himalaya and the Tibetan Plateau. The major tectonic features in this region include, from south to north, the Main Boundary Thrust (MBT), the Main Central Thrust (MCT) and the Indus–Tsangpo Suture Zone (ITSZ). All these linear tectonic features run along the Himalayan belt from west to east having a NW–NS trend in the Northwestern Himalaya, an E–W trend in the Western, Central and Eastern Himalaya, and NE–SW trends in the North-eastern Himalaya.

Most of the seismicity in the Himalayan region is concentrated along shallow, north-dipping planes, which indicate the underthrusting of the Indian plate beneath the Eurasian plate. In addition to four great earthquakes of magnitude exceeding 8 during 1897, 1905, 1934 and 1950, another 10 earthquakes exceeding magnitude 7.5 have occurred in the Himalayan belt during the past 100 years. From a very simple consideration the whole Himalayan belt, from west to east, can be considered as one seismic source zone, but dividing the zone into smaller segments would be more appropriate. We have, therefore, segmented this region by source numbers 25 and 86 in the eastern sector; sources 34, 35 and 36 in the central sector; 40 and 41 in the western sector; and 37, 38, 39, 42, 48, 49 and 50 in the North-western sector.

In contrast to the thrusting in the Himalaya, the extraordinarily thick crust of the Tibetan plateau in the north is characterized by crustal extension and eastward extrusion, manifested by earthquakes indicating normal faulting and strike-slip motion. The Altyn Tagh, the Kunlun and the Xianshuihe are the three major fault systems supporting the left-lateral strike-slip motion (Molnar, 1992). Sources 22, 23, 24, 26, 27, 28, 29, 30, 31, 32 and 33 have been delineated in this zone, based primarily on seismicity trends (Plate 4).

The Burma–Andaman Arc marks the eastern margin of the Indian plate, along which an oblique convergence between the Indian and the Burmese plates has been suggested (Fitch, 1972; Curray *et al.*, 1979). The major tectonic features along the arc are the N–S trending Indo-Burman ranges (IBR) in the north and the Andaman–Nicobar ridge (ANR) in the south. The Sumatran fault system in the sout-heast, the Western Andaman fault (WAF) and the Sagaing fault further east, are the features supporting major right lateral movements in this region. The distribution of earthquakes in the Burmese arc region suggests the presence of a subducted Indian lithospheric slab beneath the Burmese plate (Verma *et al.*, 1976; Mukhopadhyay and Das Gupta, 1988; Ni *et al.*, 1989; Gupta *et al.*, 1990). The occurrence of shallow and intermediate focus earthquakes has been reported in the Naga Hills (NH), and the Arakan Yoma fold belt (Kumar *et al.*, 1996).

While the seismic source zones 4, 5, 6, 7 and 8 cover the Burmese arc and adjoining region on the eastern side, the source zone 3 on the west encompasses the Shillong Plateau (SP), which experienced a strong earthquake of magnitude 8.7 in 1897. The tectonics of the Shillong Plateau are distinctly different from that of the regions to its north, south and west (Rao and Kumar, 1997), and hence a separate zone for the region The area further east, namely the south China region, is characterized by many faults and lineaments such as the Red River Fault Zone (RRFZ). The seismic sources 9 through 23 have been identified in this seismic province.

The nature of convergence varies from a continental type in the Burmese arc to an oceanic type in the Andaman arc, with a relatively quiet seismic zone marking the transition (Chandra, 1984, Kumar *et al.*, 1996). Shallow and occasional intermediate-depth earthquakes delineate the subducted slab under the Andaman–Nicobar islands joining the seismicity trend of the Indo-Burman ranges. A distinct, separate lineation of shallow focus earthquakes passes under the central basin of the Andaman Sea and indistinctly follows the line of the Sagaing Fault, towards the eastern Himalayan syntaxis (Curray, *et al.*, 1979). The transition seismic zone, referred to above, has been divided into sources 79 and 80. In the region south of this, along the Andaman–Nicobar region, sources 81, 82 and 83 have been assigned. Further south, the sources 84 and 85 cover the northern Sumatra region.

The North-western Himalayan region also has the characteristic Himalayan tectonic features—namely the MBT, MCT, and is bordered by the Hindukush syntaxis and the Pamir knot region in the extreme north-west. The Hindukush and Pamir knot region are characterized by the junction of several tectonic features. This plate boundary region experiences a high level of seismicity varying from shallow to intermediate-depth earthquakes. Sources 43, 44, 45, and 46 cover this region. The other prominent tectonic features in the north-western India region are the transverse fault systems known as the Chaman fault, and the Kirthar and the Sulaiman Ranges (SR). The tectonics of the Kirthar–Sulaiman ranges are influenced by transcurrent faulting. This region also belongs to the plate boundary and experiences a high level of seismicity. These areas are covered by sources 47, 51, 52, 53, 54 and 55.

Indian Shield Region

The Indian Shield region is marked by several rift zones and shear/thrust zones. Although it is considered to be a stable continental region (SCR), this region has experienced many earthquakes of magnitude 6.0 since the 18th Century (Ramalingeswara Rao, 1998), some of which were disastrous. Among them are the Mahabaleshwar (1764), Kutch (1819), Damooh Hills (near Jabalpur,

1846), Mount Abu (1848), Coimbatore (1900), Son Valley (1927), Satpura (1938), and the Jabalpur earthquake of May, 1997. The spatial distribution of these high magnitude earthquakes is notable in Plate 5. The seismicity in the shield region is quite diffused, except for a few localized alignments, which are discussed latter. Generally speaking, the Indian Shield region can be considered as one single seismic source zone for hazard computations. However, smaller seismic zones can be delineated in this region, primarily based on the locales of the major earthquakes and seismic lineaments, some of which are not so well defined. The nature of seismicity, tectonic trends, and the distribution of seismic sources is described below.

The Narmada–Son lineament (NSL) is a prominent tectonic feature of the Indian Shield, trending ENE-WSW from 21°N, 72°E on the western coast of India to 24°N, 88°E on the eastern side. This apparently divides the shield into 2 sectors, which we name as the northern and the southern sectors, to facilitate further description and discussions. The Narmada–Son region has been experiencing earthquakes of different magnitudes in the past, a recent example being the May 21, 1997 Jabalpur earthquake of magnitude 6.0. This is an SCR earthquake with an unusual focal depth of about 30 km. At least 4 earthquakes of magnitude > 5.4 have earlier occurred along this zone, two of them in the proximity of the 1997 Jabalpur earthquake (Gupta *et al.*, 1997b). We have assigned 2 probable seismic zones in the central part of the Narmada–Son lineament, numbered 63 and 64. Further, while the north-east portion of the Indian Shield is covered by sources 66, 67, 68, the region west of the Burmese arc and south of the Shillong plateau region is described by sources 1 and 2.

The major tectonic constituents in the southern sector of the Indian Shield include the massive Deccan Volcanic Province (DVP), the South Indian Granulite Terrain (SIGT), the Dharwar Craton (DC), the Cuddapah Basin (CB), the Godavari Graben (GG), the Mahanadi Graben (MG), and the Eastern and the Western Ghats on the east and west coast of India, respectively.

The Eastern Ghat region, in general, is a quiet an exceptional zone, characterized by diffused, low magnitude, shallow focus earthquakes and an occasional earthquake of magnitude 5-6. The preferred fault plane solutions generally indicate a northeast–southwest orientation with left lateral strike slip motion. An alternate set of solutions indicate thrust faulting along the north-west orientation. Not very many historical earthquakes are reported to have occurred in this region. Based on localized concentration of seismicity, we have delineated sources 70, 71 and 74 along the Eastern Ghat region.

The Western Ghat region of the Indian Shield also depicts diffused seismicity, except for some clusters, the prominent one being the Koyna–Warna region. The Koyna reservoir region has been experiencing induced earthquakes right from the date of its first filling in 1962. Over the past 34 years, the region around the reservoir has experienced 10 earthquakes of magnitude ≥ 5, over 100 earthquakes of magnitude ≥ 4, and about 100,000 of smaller magnitudes. The world's largest reservoir-induced earthquake of magnitude 6.3 occurred in the Koyna region on December 10, 1967. The Warna reservoir, located 20 km south of Koyna, began to be filled in 1986. The Koyna–Warna region has been experiencing a burst of seismicity since August 1993. More than a thousand earthquakes of magnitude around 2–3 have occurred since then. Two earthquakes of magnitude 5 and 5.4 occurred on December 8, 1993 and February, 1994, respectively. Gupta *et al.* (1997a) give a detailed picture of seismicity in the Koyna–Warna region. The seismicity in the region aligns itself beautifully along the local/regional fault systems. Thus, the Koyna–Warna region does assumes a special relevance from the seismic hazard point of view. This region, therefore, has been considered as a separate seismic source zone numbered 77. The region along the western coast, north of Koyna, has been considered as two sources, 78 and 62.

The source 78 around Bombay is characterized by N–S tectonic lineaments, such as the Panvel Flexure (PF) and similarly aligned seismicity patterns. This source region is also reported to have experienced a large earthquake in the historic times. The source 62 encompasses the NW–SW trending features in the Cambay basin and the western end of the W–E trending Narmada–Son lineament also depicts a reasonable level of seismicity, enough to merit designation of an independent seismic source. Down south, near Trivandrum, along the western margin, we have delineated source 72, primarily on account of some recent concentration of seismic activity.

In the central shield region south of the Narmada–Son lineament, the seismic activity is considerable, although diffused. The Latur region in central India experienced an earthquake of Mw 6.1 in 1993. The inferred depth was about 5 km and the focal mechanism by several agencies indicated a thrust-type faulting with the P axis trending NNE, consistent with the direction of the India–Eurasia plate motion. The inferred fault plane was NW striking and dipping 45 degrees SW. Seeber *et al.* (1996) suggested that the earthquake was caused by a new fault in the Deccan trap region. On the other hand, Rajendran *et al.* (1996) inferred the trend of a pre-existing fault based on a study of Landsat images prior to the earthquake. On the basis of borehole studies across the fault, which indicate large displacement of 3 to 6 meters, Gupta *et al.* (1998) suggested that the displacement is far too much to be accounted by a single earthquake of Mw 6.1 and concluded that the fault is a pre-existing one. It appears that this region has also been active in the historical times. In view of these considerations, the region around Latur has been taken as a seismic source zone numbered 76. The source 69 covers the

Godavari Graben region which experienced a moderate sized earthquake of M 5.3, known as the Bhadrachalam earthquake, in the year 1969. The region around Bellary and Coimbatore have been demarcated as source zones 75 and 73, respectively, on account of having experienced moderate-sized earthquakes in the past, as mentioned in the earlier section.

The northern sector of the Indian Shield has relatively lower levels of seismicity. The region has a prominent tectonic feature called the Bundelkhand Craton (BC) in the central area, bounded by NNE–SSW to N–S trending lineaments on the west as well as on the east. Sources 57, 60, 61 and 65 are delineated in this region.

The north-western corner of Indian Shield (in the vicinity of 24°N and 72°E) is characterised by N–S, NW–SE and E–W tectonic trends and shows relatively higher level of seismicity. Sources 56, 58 and 59 have been identified in this region. The source zone 59 experienced the well-known 7.8 magnitude Kutch earthquake of 1819.

Source Zone Characterization

To facilitate the steps for source zone characterization, a software toolbox was developed, which performs the essential data handling and pre-processing tasks like:

1. scanning earthquake catalogues (of different formats) to segregate events for the specified rectangular block defined by latitude–longitude limits, or a polygonal block defined by latitudes and longitudes of the vertices,
2. merging different catalogues and sorting in chronological order,
3. removal of the duplicates,
4. removal of aftershocks according to a defined criteria and preparation of a main-shock catalogue,
5. plotting of epicenters by superposing the source zones and tectonic features, geographical locations, etc.,
6. estimation of ' a' and 'b' values, as well as
7. extracting required information from the output files of FRRISK88M, for the preparation of hazard maps.

The minimum magnitude was assigned as 5.0 for all the source zones, because it is the lower level of magnitude that would cause a hazard of a significant level. The maximum magnitude was estimated from the past seismicity for each of the zones separately. The seismic parameters 'a' and 'b' were estimated by applying the maximum likelihood method and converted to Nu and Beta, which go as inputs to the hazard computations.

The source zones corresponding to the seismically active regions in the plate boundary and adjacent regions contain enough statistics for an independent computation of 'a' and 'b' values. However, the source zones in the Indian Shield do not contain sufficient information for this purpose. To overcome this problem, the 'b' value was computed for the whole shield region as one unit and was assigned to each zone within it, while the 'a' value was computed for each zone separately. A similar logic was adopted for a few other source zones with sparse seismicity. The details of the characteristics of each source zone are listed in Table 1.

Table 1 *The characteristics of the seismic source zones*

Source No.	Max. Mag.	Nu (mag.5.0)	Beta	Source No.	Max. Mag.	Nu (mag.5.0)	Beta
1	6.5	0.016	1.379	29	8.0	0.426	0.997
2	8.5	0.378	1.194	30	7.0	0.739	1.308
3	8.7	0.422	1.449	31	6.5	0.247	1.394
4	8.5	0.566	1.134	32	7.0	0.164	1.394
5	8.5	1.712	1.134	33	7.0	0.077	1.394
6	8.0	0.482	1.570	34	8.0	0.171	1.230
7	8.0	0.201	1.130	35	8.0	0.631	1.406
8	8.0	0.591	1.340	36	7.5	0.086	1.406
9	7.0	0.416	1.008	37	7.0	0.143	1.634
10	7.0	0.602	1.068	38	7.0	0.030	1.634
11	7.0	0.172	1.263	39	7.0	0.174	1.764
12	7.5	0.243	1.062	40	7.5	0.155	1.193
13	6.5	0.352	1.139	41	8.5	0.405	1.388
14	7.5	0.410	1.445	42	7.5	0.298	1.323
15	7.5	0.287	1.268	43	7.5	0.069	2.131
16	8.5	0.353	1.071	44	8.0	3.385	1.610
17	7.0	0.383	1.278	45	8.5	1.434	1.549
18	7.5	0.192	1.387	46	7.5	0.042	1.549
19	7.5	0.221	1.219	47	7.5	0.689	1.617
20	8.0	0.285	1.219	48	7.0	0.150	1.649
21	7.5	0.487	1.292	49	7.5	0.089	1.671
22	8.0	0.199	1.421	50	7.0	0.104	1.671
23	7.0	0.422	1.432	51	7.0	0.151	1.919
24	8.5	0.704	1.253	52	8.0	0.205	1.771
25	8.5	0.864	1.073	53	7.0	0.079	1.634
26	8.0	0.270	0.979	54	7.5	0.349	1.668
27	8.5	0.133	1.495	55	7.5	0.154	1.634
28	8.0	0.494	1.362	56	7.0	0.074	1.795

Continues...

Table 1 *Continued...*

Source No.	Max. Mag.	Nu (mag.5.0)	Beta	Source No.	Max. Mag.	Nu (mag.5.0)	Beta
57	7.0	0.001	1.379	72	6.0	0.008	1.379
58	7.5	0.011	1.266	73	6.0	0.016	1.379
59	8.5	0.126	1.266	74	6.0	0.001	1.379
60	6.5	0.034	1.379	75	6.0	0.006	1.379
61	7.5	0.026	1.379	76	6.5	0.003	1.379
62	6.5	0.049	1.379	77	6.5	0.128	1.379
63	6.5	0.034	1.379	78	6.5	0.128	1.379
64	6.5	0.022	1.379	79	6.5	0.148	1.378
65	6.0	0.008	1.379	80	7.5	0.139	1.378
66	6.5	0.001	1.379	81	8.5	0.694	1.434
67	6.0	0.039	1.379	82	7.0	0.677	1.367
68	6.5	0.028	1.379	83	7.5	1.809	1.361
69	6.5	0.033	1.379	84	7.5	1.676	1.440
70	6.0	0.050	1.379	85	7.5	1.259	1.409
71	6.0	0.006	1.379	86	8.5	0.395	1.495

The Seismic Hazard Map

Using the probabilistic hazard assessment approach of McGuire adopted by GSHAP, the Peak Ground Accelerations (PGA) were computed using the FRISK88M software for 10% probability of exceedence in 50 years, at locations defined by a grid of 0.5° × 0.5° in the region 0°N–40°N and 65°E–100°E. Since no reliable estimates of attenuation values are available for the Indian region, the attenuation relation of Joyner and Boore (1981) was used. The PGA values over the grid were contoured to obtain a seismic hazard map (Plate 6). A contour interval of 0.05g was chosen, since we believe that problems and uncertainties associated with the source zone definition and paucity of the seismicity information would not permit better resolution than 0.05g. The hazard map depicts that a majority of the plate boundary region and the Tibetan Plateau region have hazard levels of the order of 0.25g, with prominent highs of the order of 0.35–0.4g in the seismically active zones like the Burmese Arc, North-eastern India and North-west Himalaya/Hindukush region. In the Indian Shield region, the regional seismic hazard covering a major area is of the order of 0.1g, whereas some locales like Koyna depict hazard to the level of 0.20g.

Concluding Remarks

The computational experience in the GSHAP exercises has brought out certain concerns. The major concerns that have emerged are that the Indian region poses a lot of problems for the data homogenization efforts, and the seismic hazard maps are strongly influenced by the size and shape of seismic source zones. Another key issue, which is a cause of concern is the non-availability of representative strong motion attenuation formulae, which compelled us to be satisfied with applying one of the internationally developed formulae. The seismic hazard map of India and adjoining regions, presented here, is one possible map with all the constraints outlined. This map, we believe, gives a reasonably representative seismic hazard picture over the region covered.

We have carried out computations considering the probable source zone depth as 10 km. Any depth shallower than this would be unrealistic, since large earthquakes do not take place at very shallow depths. Any depth deeper than 10 km would also not have much meaning, because it would reduce the hazard considerably and one needs to adopt a conservative but realistic approach. The hazard values in the map, presented here, are lower than the ones given by Khattri *et al.* (1984). We would get same level of hazard if the source depth was put at 0 km. Apparently, Khattri *et al.* (1984) took a source depth of 0 km, which produced higher values of hazard estimates.

ACKNOWLEDGMENTS

The authors are thankful to Dr H K Gupta, for his support and guidance, Dr N Puranchandra Rao for his help in compiling the tectonic map and to Dr B R Rao for providing his unpublished catalogue of the Indian Shield region. The help rendered by Mrs M Murli Kumari at all the stages of data preparation, data processing and report preparation is thankfully acknowledged. Sincere thanks are also expressed to Dr Peizhen Zhang for discussions and collaborative work in the test area.

REFERENCES

Bhatia S C, Gupta, H K, Ravi Kumar, M, Rao, N P, Chitrakar, G R, Zhang, P and Yang, Z-X (1997) Seismic hazard map of GSHAP Area XII comprising eastern Himalaya, Northeastern India, Burmese arc, South China and adjoining regions, *Abstract, 29th General assembly of IASPEI, August 18–28, 1997*, Thessaloniki, Greece, p. 387.

Chandra (1984) Tectonic Segmentation of the Burmese–Indonesian Arc, *Tectonophysics*, **105**: 279–290.

Curray, J R, Emmel, F J, Moore, D G, Raitt, R W, Henry, M, and Kieckhefer, R (1979) Tectonics of the Andaman sea and Burma, In: Watkins, J S, Montadert, L and Dickerson, P (Eds), *Geological and Geophysical Investigations of Continental Margins, AAPG Memoir*, **29**: 189–198.

Fitch, T J (1972) Plate convergence, transcurrent faults and internal deformation adjacent to SE Asia and Western Pacific, *J. Geophys. Res.*, **77**: 4432–4460.

Gupta, H K, Rastogi, B K and Narain, H (1972a) Common features of the reservoir associated seismic activities, *Bull. Seismol. Soc. Am.*, **62**: 481–492.

Gupta, H K, Rastogi, B K and Narain, H (1972b) Some discriminatory characteristics of earthquakes near Kariba, Kremasta and Koyna artificial lakes, *Bull. Seismol. Soc. Am.*, **62**: 493–507.

Gupta, H K, Fleitout, L, and Froidevaux, C (1990) Lithospheric subduction beneath the Arakan Yoma fold belt: Quantitative estimates using gravimetric and seismic data, *J. Geol. Soc. India*, **35**: 235–250.

Gupta, H K, Kanal, K N, Upadhyay, S K, Sarkar, D, Rastogi, B K, and Duda, S J (1995) Verification of magnitudes of Himalayan region earthquakes of 1903–1985 from Gottingen Observatory, *Tectonophysics*, **244**: 267–284.

Gupta, H K, Rastogi, B K, Chadha, R K, Mandal, P, and Sarma, C S P (1997a) Enhanced reservoir induced earthquakes in Koyna region, India, during 1993–95, *Jour. Seismol.*, **1**: 47–53.

Gupta, H K, Chadha, R K, Rao, M N, Narayana, B L, Mandal, P, Ravi Kumar, M and Kumar, N (1997b) The Jabalpur earthquake of May 22, 1997, *Jour. Geol. Soc. India*, **50**: 85–91.

Gupta, H K, Dwivedi, K K, Banerji, D C, Rao, R U M, Rao, G V, and Srinivasan, R, (1998) Latur earthquake, Maharashtra India: A case study of a borehole investigation at SCR earthquake site, *Abstract, Chapman Conference on Stable Continental Region Earthquakes, January 25–29, 1998*, NGRI, Hyderabad, India, p. 39.

Joyner, W B and Boore, D M (1981) Peak acceeleration and velocity from strong-motion records including records from the 1979 Imperial Valley, California, Earthquake, *Bull. Seismol. Soc. Am.*, **71**: 2011–2038.

Khattri, K N, Rogers, A M, Perkins, D M and Algermissen, S T (1984) A seismic hazard map of India and adjacent areas. *Tectonophysics*, **108**: 93–134.

Khattri, K N (1987) Great earthquakes, seismicity gaps and potential for earthquake disaster along the Himalaya plate boundary. In: Mogi, K, and Khattri, K N (Eds), Earthquake Prediction, *Tectonophysics*, **138**: 79–92.

Leloup, P H, Lacassin, R, Tapponier, P, Schaerer, U, Dalai, Z, Xiaohan, L, Lianshang, Z, Shaocheng, J, Trong and Trinh, P T (1995) The Ailao Shan-Red

River shear zone (Yunnan, China), Tertiary transform boundary of Indochina, *Tectonophysics*, **251**: 3–84.

Molnar, P (1992) A review of seismicity, recent faulting and active deformation of the Tibetan plateau, *J. Him. Geol.*, **3**: 43–78.

Molnar, P, Fitch, T S, Wu F T (1973) Fault plane solutions of shallow earthquakes and contemporary tectonics in Asia, *Earth Planet. Sci. Lett.*, **19**: 101–112.

Mukhopadhyay, M, and Das Gupta, S (1988) Deep structure and tectonics of the Burmese arc: constraints from earthquake and gravity data. *Tectonophysics*, **149**: 299–322.

Ni, J F, Speziale, M G, Bevis, M, Holt, W E, Wallace, T C and Seager, W R (1989) Accretionary tectonics of Burma and the three dimensional geometry of the Burma subduction. *Geology*, **17**: 68–71.

Rajendran, C P, Rajendran, K and John, B (1996) The 1993 Killari (Latur) Central India earthquake: An example of fault reactivation in Precambrian crust, *Geology*, **24**(7): 651–654.

Ramalingeswara Rao, B (1998) A Catalogue of Indian Shield Earthquakes. Personal Communication.

Ravi Kumar, M and Rao, N P (1995) Significant trends related to the slab seismicity and tectonics in the Burmese arc region from Harvard CMT solutions, *Phys. Earth and Planet. Inter.*, **90**: 75–80.

Ravi Kumar, M, Rao, N P and Chalam, S V (1996) A Seismotectonic study of the Burma and Andaman arc regions using Centroid Moment Tensor data. *Tectonophysics*, **253**: 155–165.

Ravi Kumar, M, Rao, N P, and Bhatia, S C (1998) Analysis of Harvard CMT solutions and computation of strain rates in the Himalaya and Tibetan plateau regions, *Current Science*, **75**: 311–316.

Rao, N P and Ravi Kumar, M (1997) Uplift and tectonics of the Shillong plateau, northeast India, *J. Phys. Earth*, **45**(3): 167–176.

Seeber, L, Ekstrom, G, Jain, S K, Murthy, C V R, Chandak, N and Ambruster, J G (1996) The Killari earthquake in central India: A new fault in Mesozoic basalt flows? *Jour. Geophys. Res.*, **101**(B4): 8543–8560.

Tandon, A N (1956) Zones of India liable to earthquake damage, *Ind. J. Meteorol. Geophys.*, **10**: 137–146.

Verma, R K, Mukhopadhyay, M and Ahluvalia, M S (1976) Earthquake mechanisms and tectonic features of northern Burma, *Tectonophysics*, **32**: 387–399.

Major and Great Earthquakes in the Himalayan Region: An Overview

Harsh K Gupta*

Abstract

The Himalayan portion of the Alpide belt is seismically one of the most active intra-continental region anywhere in the world. Between 1897 and 1952 there was a phase of very high seismic activity when 14 major earthquakes ($M \geq 7.5$), including 5 great earthquakes of $M \geq 8$, occurred. No major earthquake has occurred since 1952. A re-estimation of the magnitude of the earthquakes in the early part of 20th century and a comparison of various catalogues has confirmed that the quiescence of a major earthquake since 1952 is real. There have been efforts to estimate the repeat time of earthquakes in the vicinity of the Himalayan region. However, most of these were statistical and related approaches, using earthquake catalogue, and information of the historical earthquakes. An effort to date repeat time on the Chedrang Fault—the locale of the 1897 Shillong Plateau earthquake of $M \geq 8.7$ has revealed that in the past at least 4 comparable earthquakes had occurred in this region, with a repeat time of about 400–600 years. In another interesting study, it has been shown that if the 1905 Kangra earthquake, which had claimed 30,000 lives was to repeat today, between 88,000 and 3,44,000 human lives would be lost—depending, of course, upon the time of the day when the earthquake occurs. This requires adequate preparation and emergency planning to reduce the seismic risk.

Keywords: Alpide Belt, earthquake risk, Himalaya, major earthquakes, repeat time.

* National Geophysical Research Institute, Hyderabad-500 007, India
Presently at
Department of Ocean Development, New Delhi 110 003, India

Introduction

The Himalayan frontal arc constitutes the central part of the Alpide seismic belt. As far as earthquakes are concerned, this is one of the most active intra-continental regions in the world. As is well-known, the northward movement of the Indian plate caused continental collision and created the Himalayan mountain range. The current seismic activity is the result of the continental collision between the Indian and the Eurasian plates. The instrumental recording in the region started towards the end of the 19th century with the installation of Milne seismograph at Alipore, Calcutta in 1898, and later at Colaba, Bombay, and Madras. Table 1 gives a list of the earthquakes of magnitude ≥ 7½ in the Himalayan region since 1897. Figure 1 depicts the earthquakes of magnitude ≥ 7, and the ones that claimed human lives at Himalayan frontal arc. This has been updated from Chandra (1992).

Table 1 *Earthquake of M ≥ 7½ in the Himalayan region since 1897*

Date	Latitude (N)	Longitude (E)	Location	Magnitude (m)
June 12, 1897	25.9	91.8	Assam	8.7
April 4, 1905	33.0	76.0	Kangra Valley	8.6
Dec. 12, 1908	26.5	97.0	N. Burma	7.5
May 23, 1912	21.0	97.0	Burma	8.0
July 8, 1918	24.5	91.0	Assam	7.6
Jan. 27, 1931	25.6	96.0	N. Burma	7.6
Jan. 12, 1934	26.5	86.5	Bihar–Nepal	8.4
May 30, 1935	29.5	66.7	Quetta	7.6
Sept. 12, 1946	23.5	96.0	India–Burma	7.7
July 29, 1947	28.5	94.0	N.E. Assam	7.9
Aug. 15, 1950	28.5	96.7	Assam	8.7
Nov. 18, 1951	31.1	91.4	North of India	8.0
Aug. 17, 1952	30.5	91.5	North of India	7.5

Quiescence of Major Earthquakes

The earthquake catalogue for the Himalaya and North-east India regions (latitude 20°N to 38°N and longitude 75°E to 100°E) comprising of portions of Himalaya and Arakan Yoma fold-belt for the period 1987 to the

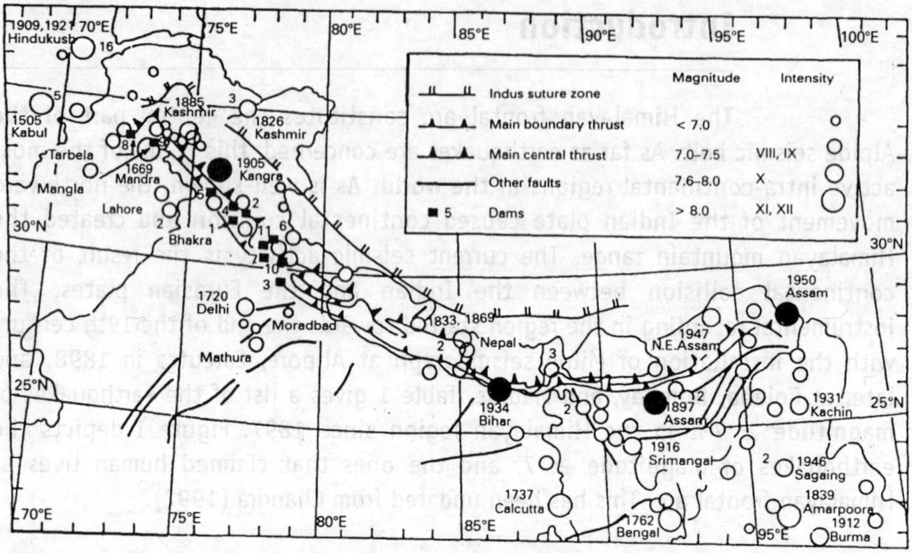

Figure 1 *Earthquakes of magnitude greater and equal to 7 as well as earthquakes that claimed human lives in the vicinity of the Himalayan Frontal Arc (updated from Chandra, 1978).*

present shows that there have been a total of 14 major earthquakes (M ≥ 7.5) during the period 1897 to 1952. This includes 5 great earthquakes exceeding magnitude 8: i.e., the Shillong Plateau earthquake of 1897, the Kangra earthquake of 1905, the Bihar–Nepal earthquake of 1934, and the India–China earthquake of 1950 and its aftershock of M 8 of November 1951. No major earthquake of magnitude 7.5 or larger has occurred since 1952. There was an apprehension that possibly during the first half of the 20th century the earthquake magnitudes were over-estimated in the Himalayan region. To verify the same, we took up a detailed study, where we redetermined the magnitudes of 37 earthquakes (M ≥ 7) in the Himalayan region for the period 1903–1985 (Gupta *et al.*, 1995). For these earthquakes, the microfilms of the Göttingen Observatory seismograms were used and the magnitude values were estimated using P-, S- and surface waves from long- and medium-period instruments. The redetermined magnitudes of the Himalayan earthquakes were within the error limits of observations of the earlier magnitude values and confirmed that, indeed, no major earthquake of M ≥ 7.5 has occurred since 1952. In Table 2, the distribution of the earthquakes in different magnitude ranges for the periods 1897 to 1952 and 1953 through to 1999 is given. (updated from Satyabala and Gupta, 1996).

Table 2 *Number of earthquakes in different magnitude ranges during the period 1897–2001 in the Himalayan region*

Magnitude	1897–1952	1953–2001
$M \geq 7.5$	14	0
$7.5 > M \geq 7.0$	11	7
$7.0 > M \geq 6.5$	19	21

An examination of Table 2 shows that while major earthquakes of $M \geq 7.5$ have not occurred since 1952, there are a comparable number of earthquakes in 6.5 to 7.4 magnitude ranges, indicating that indeed there is a quiescence of major earthquakes in the Himalayan region. This quiescence was further confirmed by comparing several other catalogues. For example, those of Abe (1981, 1984 and 1994), and Pacheco and Sykes (1992).

Return Period of Great Earthquakes

There has been a sincere effort to estimate the repeat time of earthquakes in the vicinity of the Himalayan region. One paper which needs to be mentioned is that of Seeber and Armbruster (1981) where, from a detailed study of the spatio-temporal distribution of earthquakes in the Himalayan region, they have estimated that the entire Himalayan arc ruptures in about 180 to 240 years; and the repeat time of a typical magnitude 8 earthquake with a rupture length of 300 km is 200 to 270 years. However, such efforts are hampered by the fact that our earthquake catalogues for the Himalayan region are not very reliable and, in any case, the instrumental data is available only for the period of the last 100 years; i.e., from 1898 till the present. Over the past few decades a sincere effort in different parts of the world has been made to identify, characterize, and date the deformed sedimentary structures caused by the past violent earthquakes and, thereby, estimating the time of occurrence of past earthquakes (Sieh, 1978; Yeats *et al.*, 1997 and McCalpin 1996). Sukhija *et al.* (1999) have reported the result of paleo-liquefaction evidence on the periodicity of large pre-historic earthquake in Shillong Plateau. They conducted very systematic studies by collecting samples in several trenches opened in the Krishnai and Dudhani rivers, close to the Chedrang fault in the Shillong Plateau. The Chedrang fault is known to have been associated with the great 1897 Shillong Plateau earthquake. They found very well-preserved liquefaction and deformed syndepositional features at 10 selected sites in the alluvial deposits along these rivers. Using the C^{14} dating method, they have provided evidence of events which occurred in the past. In addition to the 1897 event, they found two more events that occurred during 1450 to 1650, and 700 to 1050 AD (Table 3). The third event pre-dates 600 AD. They infer a return period of 400 to 600 years for large earthquakes in the Shillong Plateau region.

This finding indicates that earthquakes of the type of 1897 in the Shillong Plateau may repeat only after about 400 to 600 years. However, what is required is to make similar estimates at several other sites where great earthquakes are known to have occurred, such as the 1934 Bihar–Nepal earthquake, and the 1905 Kangra earthquake.

A Medium Term Earthquake Forecast in North-east India

There is a global effort to establish precursor patterns that precede major earthquakes. An effort in this direction was made by Gupta and Singh (1986, 1989) in the North-east Indian region. Encouraged by the discovery of precursory swarm and quiescence preceding the Cachar earthquake of December 30, 1984, Gupta and Singh (1986, 1989) carried out an in-depth study of all the earthquakes of magnitude ≥ 7½ since 1987 and several smaller magnitude earthquakes that occurred after 1962. In their study, they discovered that the main shock magnitude (Mm) has correspondence with the magnitude of the largest events (Mp) in the swarm, and the time interval (Tp) between the onset of swarm and the occurrence of main shock in days. The following are the relations found by them:

$$Mm = 1.37\ Mp - 1.41 \text{ and}$$

$$Mm = 3 \log Tp - 3.27$$

Gupta and Singh (1986) observed that it is important to recognize swarm and quiescence before the occurrence of the main shock. They discovered one such region in the vicinity of Indo–Burma border and concluded that: "(1) Moderate magnitude to great earthquakes in the North-East India region are found to be preceded, generally, by well-defined earthquake swarms and quiescence periods, (2) On the basis of an earthquake swarm and quiescence period, an area bound by 21°N and 25½°N latitude and 93°E and 96°E longitude is identified to be the site of a possible future earthquake of M 8 ± ½, with a focal depth of 100 ± 40 km. This earthquake should occur any time from now onwards. Should it not occur till the end of 1990, this forecast could be considered as a false alarm."

Table 3 *^{14}C ages of organic samples representing various seismic events in the Shillong Plateau (after Sukhija et al., 1999)*

Event	Sample code	Minimum age	Coeval age	Maximum age	Age-range AD (MPAE)
I	AS-1:	1430-1955	AS-33: Modern	AS-18: Modern	1897[a]
	AS-3:	1425-1955	AS-34: Modern	AS-43: Modern	
II	AS-39:	1050-1415	AS-13: 1290-1950	AS-19: 1225-1660	
			AS-31: 1240-1950	AS-28: 1290-1950	1450-1650
			AS-32: 1300-1955	AS-30: 1400-1630	
			AS-45: 1450-1660		
III	AS-7:	420-1010	AS-11: 689-1290	AS-8: 530-1295	
	AS-7a:	440-1255	RAS: 645-1017	AS-24: 790-1300	700-1050
	AS-10:	660-1280		AS-37: 550-1030	
	ASS-22:	540-890		AS-41: 1020-1460	
	AS-23:	400-1000			
IV		-	-	AS-16: 240-870	earlier than 600

MPAE = most probable age of the event. Modern = ^{14}C 50 years. RAS = Rastogi *et al.* (1993)

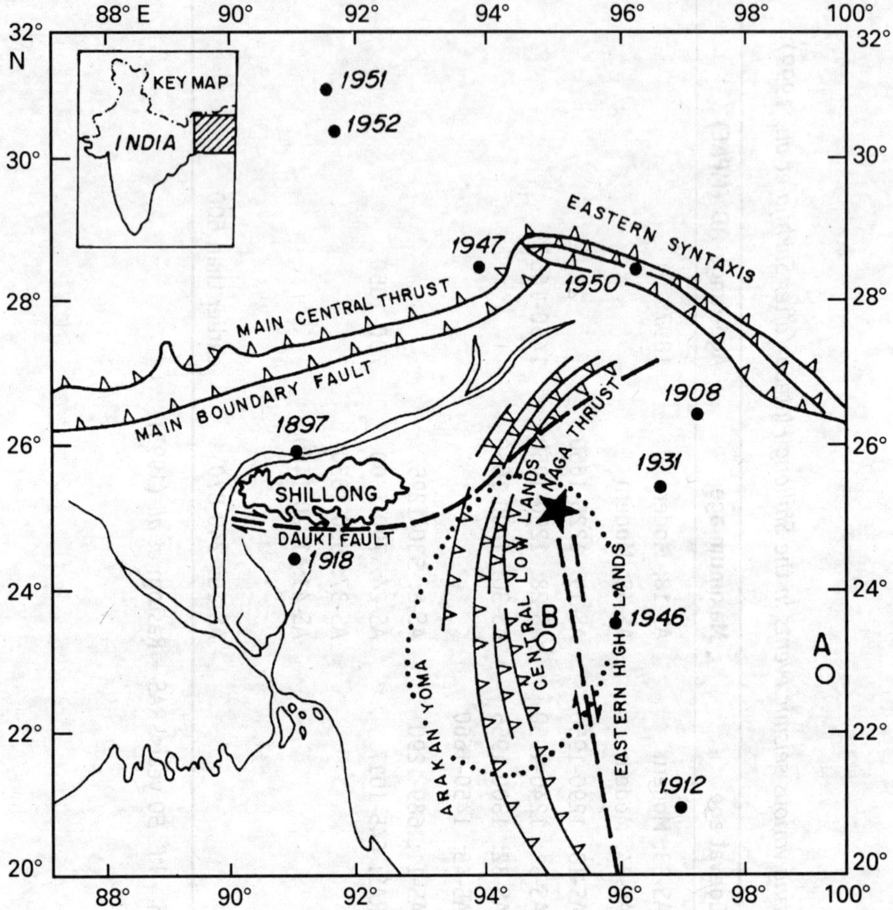

Figure 2 *Earthquakes of M ≥ 7.5 in the north-east India region since 1897 (filled circles). Elliptical area (dots) shows the preparation zone for an M = 8 ± 0.5 earthquake identified by Gupta and Singh (1986). After the last M = 7.5 earthquake of August 17, 1952 the first largest earthquake of M = 7.3 in the entire region occurred on August 6, 1988 (star). This was followed by two more earthquakes of M = 7.3 on November 6, 1988(A) and January 5, 1991(B). Figure after Gupta and Singh (1986)*

The occurrence of the August 6, 1988 earthquake (Figure 2), with focal parameters mentioned in Table 4, has proved this medium-term forecast true. This success encourages one to make similar investigations elsewhere in the Himalayan Frontal Arc, in order to concentrate hazard-related investigations in a few critical areas.

Table 4 *Forecast of August 6, 1988 earthquake*

Earthquake	Forecast	Occurrence
Parameters	(Gupta and Singh, 1986)	NEIS (preliminary determination)
Epicentre	21°N to 25½°N	25.149°N
	93°E to 96°E	95.127°E
Magnitude	8 ± ½	7.3
Depth	100 ± 40 km	90.5 km
Time	February, 1986 to December, 1986	August 6, 1988 (00.36.26.9G.C.T.)

Future Earthquakes: How Damaging will They be?

There has been a phenomenal increase in the population density in the foothills of the Himalaya during the last several decades. Therefore, the number of people likely to be affected by great earthquakes has increased considerably, as the construction continues to be of poor quality. Arya (1992) has made an estimate of the likely damage scenario in the Kangra region if the 1905 great earthquake was to repeat today. He presumed that the distribution of earthquake intensity will be similar to the 1905 earthquake, putting areas of 500 sq km, 2,400 sq km, 5000 sq km and 26,000 sq km under intensities X, IX, VIII, and VII, respectively, on the Modified Mercalli Intensity Scale. Keeping in view the building material and construction, it is estimated that a total of about 1,45,400 houses would collapse completely and another 2,67,800 houses would suffer severe damage. Arya (1992) estimates that, depending upon the time of the day when the earthquake occurs, the loss of human lives would vary between 88,000 and 3,44,000 (Table 5). He has underlined the importance of retrofitting important buildings, to reduce the hazard.

Table 5 *Estimates of human lives likely to be lost if the Kangra earthquake of 1905, which claimed 20,000 lives, were to occur today (Arya, 1992)*

Time of Occurrence	Deaths in Collapsed Houses	Deaths in Partly-collapsed Houses	Total Potential Deaths
Midnight (Sleeping)	40%	20%	3,44,000
Morning (Awake and Sleeping)	20%	10%	1,77,000
Noon Time (Out Working)	10%	5%	88,000

Concluding Remarks

From the foregoing discussion, it is obvious that major and great earthquakes have occurred in the Himalayan region and will continue to occur. However, the entire region has become much more vulnerable due to the rapid increase in population and the vulnerability of the existing structures. This is obvious from the kind of damage that the Himalayan regions suffered from during the 1991 Uttarkashi earthquake. The Uttarkashi earthquake of October 20, 1991, although of only magnitude 6.8, claimed an estimated 2000 human lives and became the most significant earthquake in the year 1991. The occurrence of a bigger earthquake will create much larger problems. What is necessary is to educate the public on what needs to be done during and after an earthquake, as well as teach them methods of simple retrofitting of non-engineered structures, so that damage due to earthquakes could be reduced considerably. Time and again, it has been demonstrated that it is NOT EARTHQUAKES, BUT BUILDINGS THAT KILL PEOPLE. Thus, it is necessary and imperative to do the following:

- The implementation of building codes of the BIS should be made mandatory.
- To retrofit important buildings situated in Zone IV and Zone V of the Zonation Map.
- About 70% of Indians live in rural areas, in houses and dwellings made without any engineering considerations. There are methods available to strengthen their dwellings by means of some simple, very inexpensive approaches. These should be popularized.
- The microzonation of important cities of the country is a must.

All this needs to be done in a short time-frame, so that people are well-prepared to face and reduce the loss of human lives and damage to property by future major earthquakes.

REFERENCES

Abe, K (1981) Magnitudes of large shallow earthquakes from 1904 to 1980, *Phys. Earth Planet. Interiors,* **27**: 72–92.

Abe, K (1984) Complements to magnitudes of large shallow earthquakes from 1904 to 1980, *Phys. Earth Planet. Interiors,* **34**: 17–23.

Abe, K (1994) Instrumental magnitudes of historical earthquakes, *Bull. Seism. Soc. Am.,* **84**: 415–425.

Arya, A S (1992) Possible effects of a major earthquake in Kangra region of Himachal Pradesh, *Current Science, Special Issue,* **62**(1 & 2): 251–256.

Chandra, U (1992) Seismotectonics of Himalaya, *Current Science, Special Issue,* **62**(1 & 2): 40–71.

Gupta, H K, Khanal, K N, Upadhyay, S K, Sarkar, D, Rastogi, B K and Duda, S J (1995) Verification of magnitudes of Himalayan region earthquakes of 1903–1985 from Goettingen Observatory, *Tectonophysics,* **244**: 267–284.

Gupta, H K and Singh, H N (1986) Seismicity of the north-east India Region, Part II: Earthquake swarms precursory to moderate magnitude to great earthquakes, *J. Geol. Soc. India,* **28**(5): 367–406.

Gupta, H K and Singh, H N (1989) Earthquake swarms precursory to moderate to great earthquakes in the Northeast India region, *Tectonophysics,* **167**: 285–298.

McCalpin, J P (Ed.) (1996) *Paleoseismology,* Academic Press, London, p. 588.

Pacheco, J F and Sykes, L R, (1992) Seismic moment catalog of large shallow earthquakes, 1900–1989, *Bull. Seism. Soc. Am.,* **82**: 1306–1347.

Rastogi B K, Chadha, R K, Rajgopalan, G (1993) Paleoseismology studies along Chedrang fault, Meghalaya, *Curr. Sci.,* **64**: 933–935.

Satyabala, S P and Gupta, H K (1996) Is the quiescence of major earthquakes (M ≥ 7.5) since 1952 in the Himalaya and Northeast India real? *Bull. of Seis. Soc. Am.,* **86**(6): 1983–1986.

Seeber, L and Armbruster, John G (1981) Great detachment earthquakes along the Himalayan arc and long-term forecasting, AGU publication, Earthquake Prediction, An international review, Maurice Ewing Series, **4**: 259–277.

Sieh, K E (1978) Prehistoric large earthquakes produced by slip on the Andreas fault at Pallett Creek, California, *J. Geophys. Res.,* **83**: 3907–3939.

Sukhija, B S, Rao, M N, Reddy, D V, Nagabhushanam, P, Hussain, S, Chadha, R K, Gupta, H K (1999) Paleoliquefaction evidence and periodicity of large prehistoric earthquakes in Shillong Plateau, India, *Earth and Planetary Science Letters*, **167**: 269–282.

Yeats, R S, Sieh, K E, Allen, C R (1997) The geology of earthquakes, Oxford University Press, New York, p. 568.

Two Great Landslide Tragedies of India

R K Bhandari*

SUMMARY

The paper spotlights two great landslide tragedies of India namely the Alaknanda Tragedy of 20th July 1970, and the Malpa Tragedy of 18th August 1998. Both were spectacular in appearance, formidable in dimension, and devastating in their consequences. Both of them created live laboratories of Nature, offering several rare messages and priceless lessons, no classrooms or textbooks could teach. No amount of simulation modeling and experimentation could have revealed what these calamities displayed on the ground. Scientists saw in them rare signatures of Nature, which, when decoded, lent significant clues to re-construct the events to better understand their mechanisms of occurrences. The eyewitness to these tragedies provided evidence, which would have otherwise been erased forever, no matter how thorough the post-disaster investigations. The paper, *inter alia,* attempts to weave the eyewitness evidence with post-disaster investigation and study.

The Alaknanda tragedy was the consequence of breach of a landslide dam at the confluence of the rivers Patalganga and Alaknanda. It has no parallel in the recorded history of the Alaknanda Valley, from whichever angle one looked at it. The scale of occurrence, the spread of damage and their impacts were all huge. It was triggered by a curious interplay of several causative factors, which combined with the extreme of violence unleashed by the Alaknanda river itself, due to the breach of the landslide dam. But for the eyewitness accounts, one would have only seen the end-face of deluge and devastation. The eyewitness observation that the landslide dam at Patalganga dramatically reduced the water flow in the river Alaknanda at Belakuchi, and that the 'fireworks in the horizon', seen by distant observers, followed the

* Director, Centre for Disaster Mitigation and Management, Chennai, India.

bursting of landslide dam, would have otherwise been missed. The dramatic reduction of water level in the river Alaknanda at Belakuchi provided a premonition to the creation of a landslide dam upstream, thereby giving the people of the area an opportunity to escape death. The visual evidence of fireworks triggered by the clash and collision between the flying boulders, upon the breach of dam, bore testimony to the extreme of violence unleashed by the bursting landslide dam.

During the Malpa rock avalanche tragedy, the eyewitness saw the upper reaches of the mountain aflame simultaneously as he heard the thunderous noise of the rock avalanche hurtling down the mountain. This observation is significant especially because of the fact that the same mountain was found to be aflame more or less the same way, also on the 4th August 1998 (on account of landsliding), but the fact was discarded by the people as a freak event. The message is thus clear. Nature has its own mysterious ways to forewarn us, but only if we have the ability to pick up the signals. The speed of the avalanche, which ranged between 10 m/s to 40 m/s, as it hurtled down the slope, would also never have been reliably estimated, but for the observed facts.

Yet another vital observation at Malpa, which would have otherwise escaped attention completely, is the occurrence of the dust storm in tandem with the rock avalanche. It was only after the eyewitness expressed the power of the dust storm that he felt on his face, that further field evidence could be seen in the form of a thin cover of dust all over the lush green slopes in the windward direction.

All the dogs at Malpa did survive the rock avalanche tragedy, as noted the next day by the people who had assembled at the Indo-Tibetan Border Post for rescue and relief operations. However, all the horses in Malpa were killed. The reason advanced was that whereas the dogs were free to escape the rock avalanche, the horses succumbed to it, because they were tethered.

The paper explicitly describes the two landslide tragedies.

The Alaknanda Tragedy of July 1970

The Setting

The area related to the Alaknanda Tragedy is shown in Plate 7. In the upper part of the Garhwal Himalaya, the river Dhauliganga, which originates from Kunlong at Nitipass (5067 m), merges into the river Alaknanda at Vishnuprayag (285.8 km; altitude 1463 m). Both the rivers are excellent examples of antecedent drainage, indicating that these rivers are older than or antecedent to the orogenic history.

The mountains of the Garhwal Himalaya lie Northwest–Southeast. High relief, steep slopes, precipitous ravines, verdant valleys, roaring streamlets, and glistening lakes provide a very colourless description of the heavenly landscape. Bhagirathi and Alaknanda are the two major rivers of the Garhwal Himalaya, both originating from the mighty Chaukhamba range of glaciers. Bhagirathi has its origin at Gangotri on the north-western face of Chaukhamba River. Alaknanda itself originates from the Satopanth Glacier at Alkapuri, on the south-eastern side of the range near Badrinath, and meets River Saraswati at Mana Pass (5610 m).

In order to fully understand the Alaknanda tragedy, it would be appropriate to revisit the event, and recount the trail of devastation unleashed by the Alaknanda floods. The Himalayan rivers and mountains are known for their extreme vagaries and the associated road networks and road users are usually their first victims. The scars of new and old landslides, the aggravating rate of erosion due to meandering rivers, the silt-swollen riverbeds, and the consequently choked bridges, the sinking roads, and the nasty road and river blockades, etc, all appear as the off-spring of these vagaries.

Like in any other hydrological network, perennial rivers like the Alaknanda are fed by a myriad of intermittent or ephemeral streams; they carve valleys proportional to their sizes. However, it is a common sight to see the river Alaknanda and its several tributaries overflow their channels in times of floods.

Before the confluence point, the River Alaknanda from Mana (3021 m) to Joshimath (1095 m), and the River Dauliganga, from Jamuo (2621 m) to Joshimath (1905 m), both display numerous old and new landslides. The Alaknanda generally flows parallel to the mountains, but at some places it bends acutely, scooping out deep gorges. Birahiganga, Patalganga, Nandakini, Pinder and Mandakini are the tributaries of the River Alaknanda. The Girithiganga and Rishiganga are the tributaries of the Dhauliganga. The Birahiganga originates from the Nanda Ghungti Glacier. The important tributaries of the Birahiganga include two snow-fed streams from the east. The other streams are the Bhadaligadi, Rogilagad, Shyamgad, Puigadhera from the north, and Gudiyar, Gadhera and Begarnala from the south. The obliterated Gohna Lake lies near the confluence with the tributary Puigadhera.

Downstream of Vishnuprayag, right from Joshimath to Srinagar, the River Alaknanda and its tributaries negotiate deep gorges, steep gradients, and sharp bends. The Dhauliganga is perhaps the steepest at a 7.5% mean gradient. The Nandakini and Mandakini both display more or less equal mean gradients of about 6.6%, as against the mean of 4.8% for the River Alaknanda. The entire river system plays havoc with the slopes that support roads, bridges and human habitat, especially when provoked by high intensity, short duration rainfalls and cloudbursts.

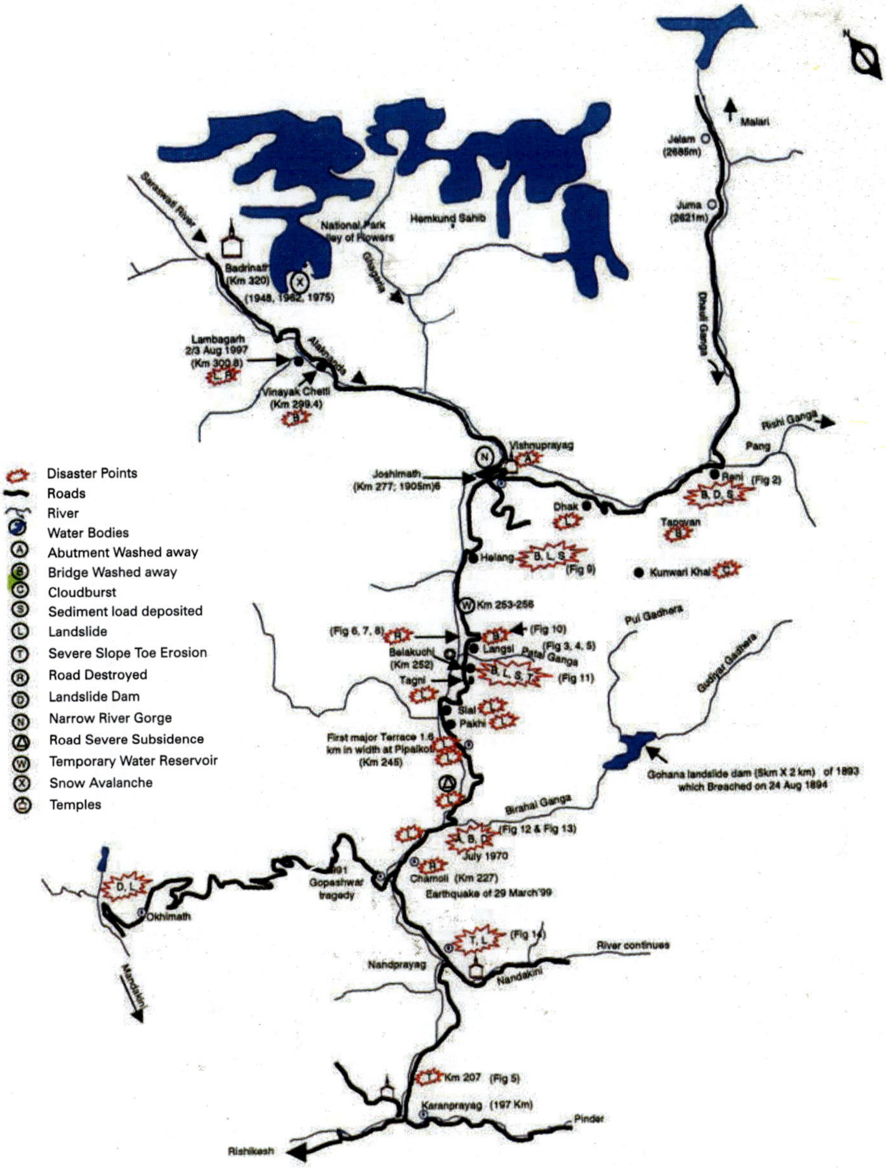

Plate 7 The trail of devastation unleashed by the landslides and the Alaknanda Floods of July 1970.

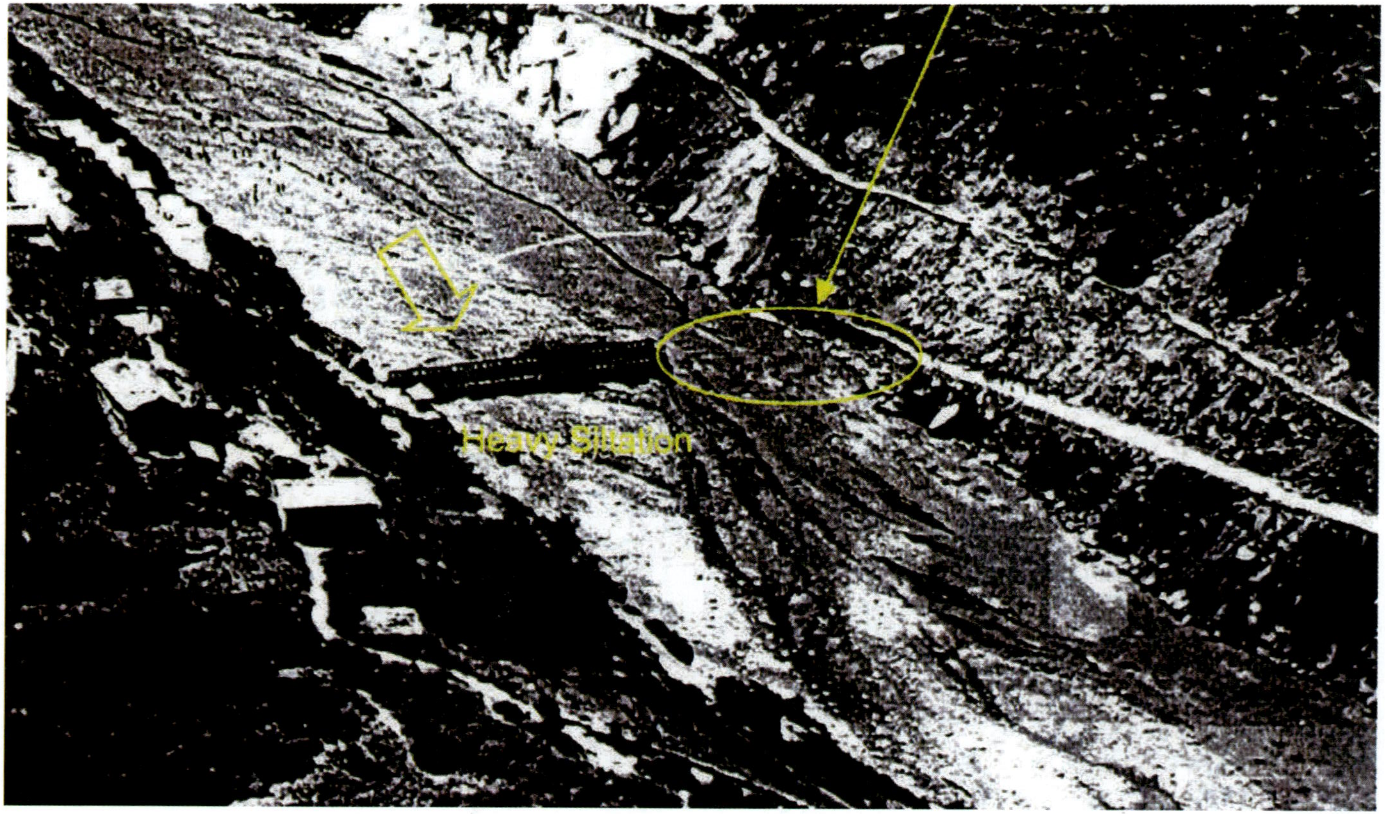

Plate 8 The left bank approaches of the Reni Bridge on river Rishi Gang at 22 km on Joshimath-Malari road was destroyed. Heavy siltation can be seen under the bridge.

Plate 9 The narrow gorge through which a colossal amount of debris was brought by Patal Ganga, a tributary of River Alaknanda, thereby leading to the 'Alaknanda Tragedy'.

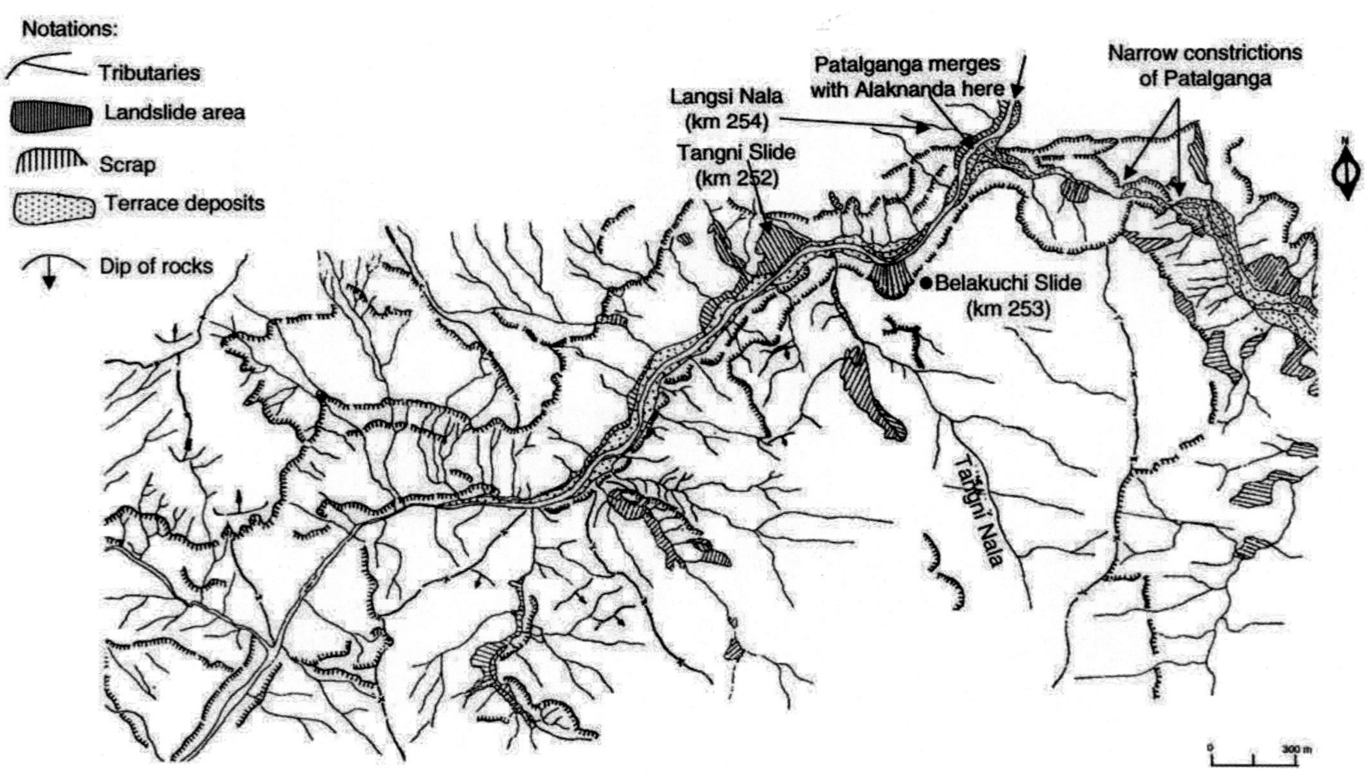

Plate 10 Landslide and slope failures along the Patalganga and Alaknanda rivers. The narrow construction of Patalganga and the associated sediment load, were responsible for the formation of the landslide dam, which burst on 20th July 1970 at 6.45 pm to send a flood wave in the River Alaknanda.

(1) Alaknanda (2) Patalganga (3) Siltation
☆ road from Langsi bridge to Joshimath
(Photograph taken from lower Yatri track)

(1) Alaknanda (2) Patalganga (3)

Plate 11 Confluence of (1) Alaknanda and (2) Ganga, at 253 m on the Rishikesh-Joshimath Road

Plate 12 River Alaknanda: (1) has thrown its right bank because of the heavy discharge of sediments (2) by river Patalganga. A thick layer of sediments (3) was seen on the road after the floodwaters receded.

Plate 13 Destruction of a bridge on the Helang Nallah. Remains of both the abutments of the bridge (R), (L) could be seen. The height of the bridge from the water level was originally about 10 m. However, the Nallah got completely silted up and consequently its bed level rose more or less to the bridge-deck level.

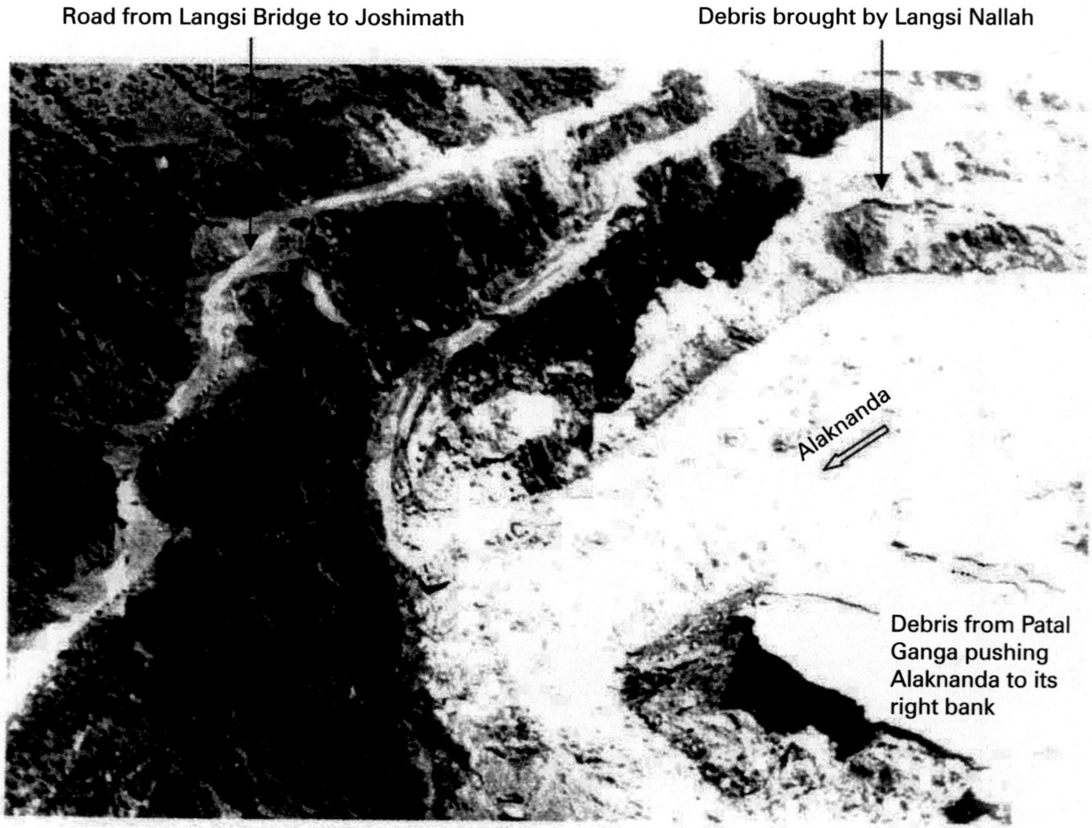

Plate 14 Langsi bridge site on the River Alaknanda.

Plate 15 Belakuchi Landslide Disaster (1) at 250 km on the (2) Rishikesh-Joshimath road. One bulldozer and one truck were trapped here at the time of disaster. (3) Heavy subsidence of the road is clearly discernible.

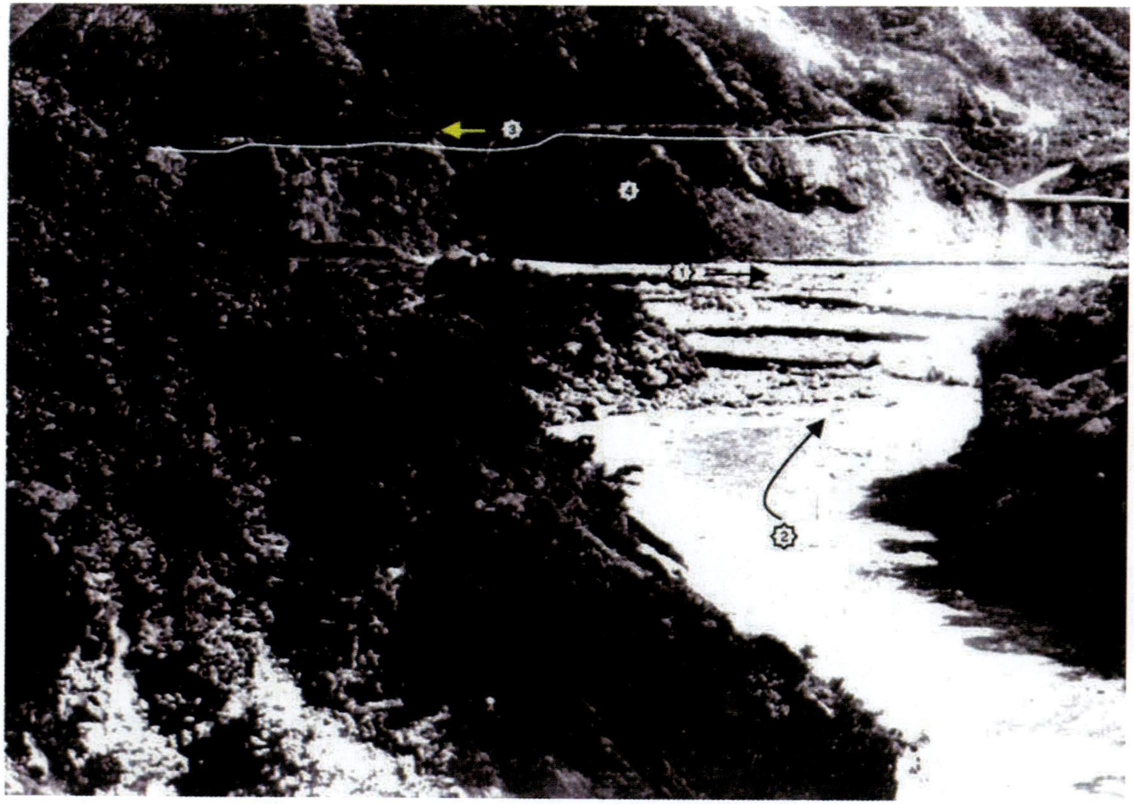

Plate 16 Confluence of river Birahi Ganga (1) with River Alaknanda (2) at Birahi. The road from Chamoli to Birahi (3) is seen in the background. Damages to the road formation and its slopes (4) are clearly visible.

Plate 17 A view of the bridge at Birahi. (1) The left bank has been completely washed away.
(Photos taken from the downstream right bank)

Plate 18 Heavy toe erosion (1) due to the River Alaknanda (2) at Nandprayag.

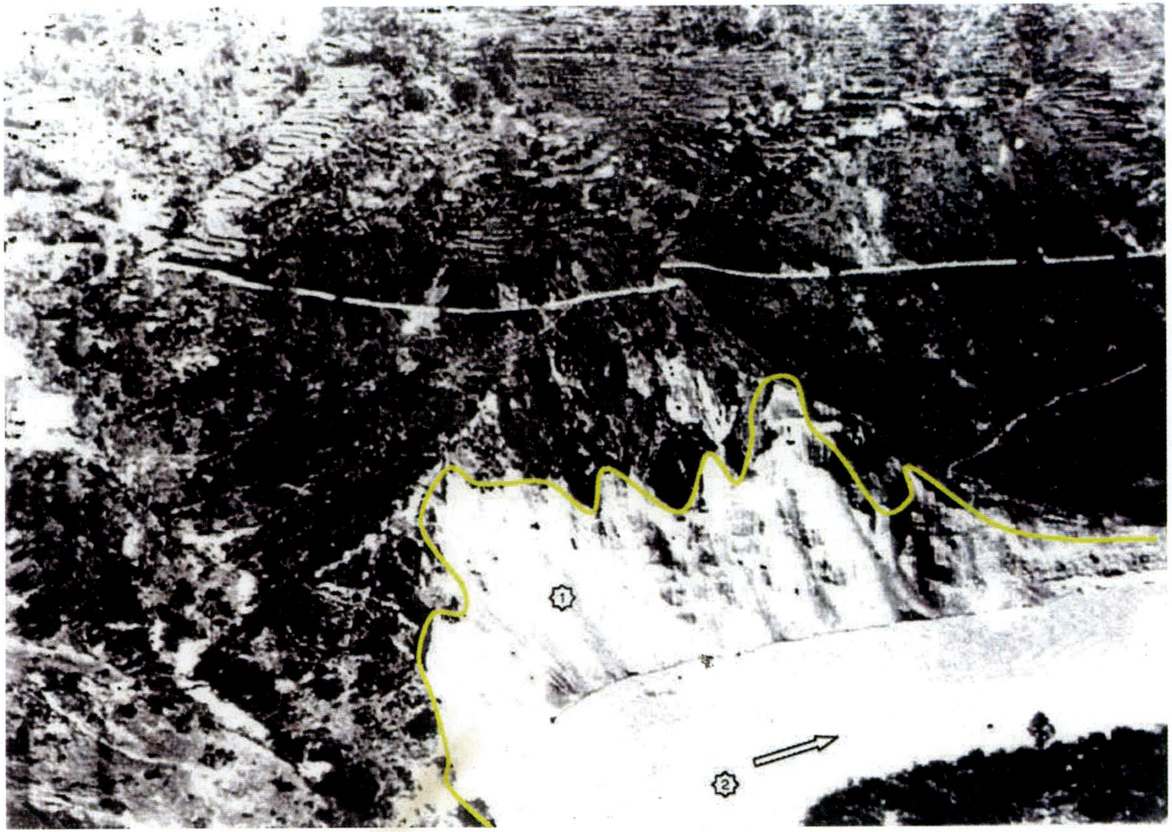

Plate 19 A helicopter photo of the Rishikesh-Joshimath road showing (1) toe erosion of slopes on the left bank of the River Alaknanda (2) at 207 Km.

Plate 20 The Yatripath after the Alaknanda tragedy of July 1970.

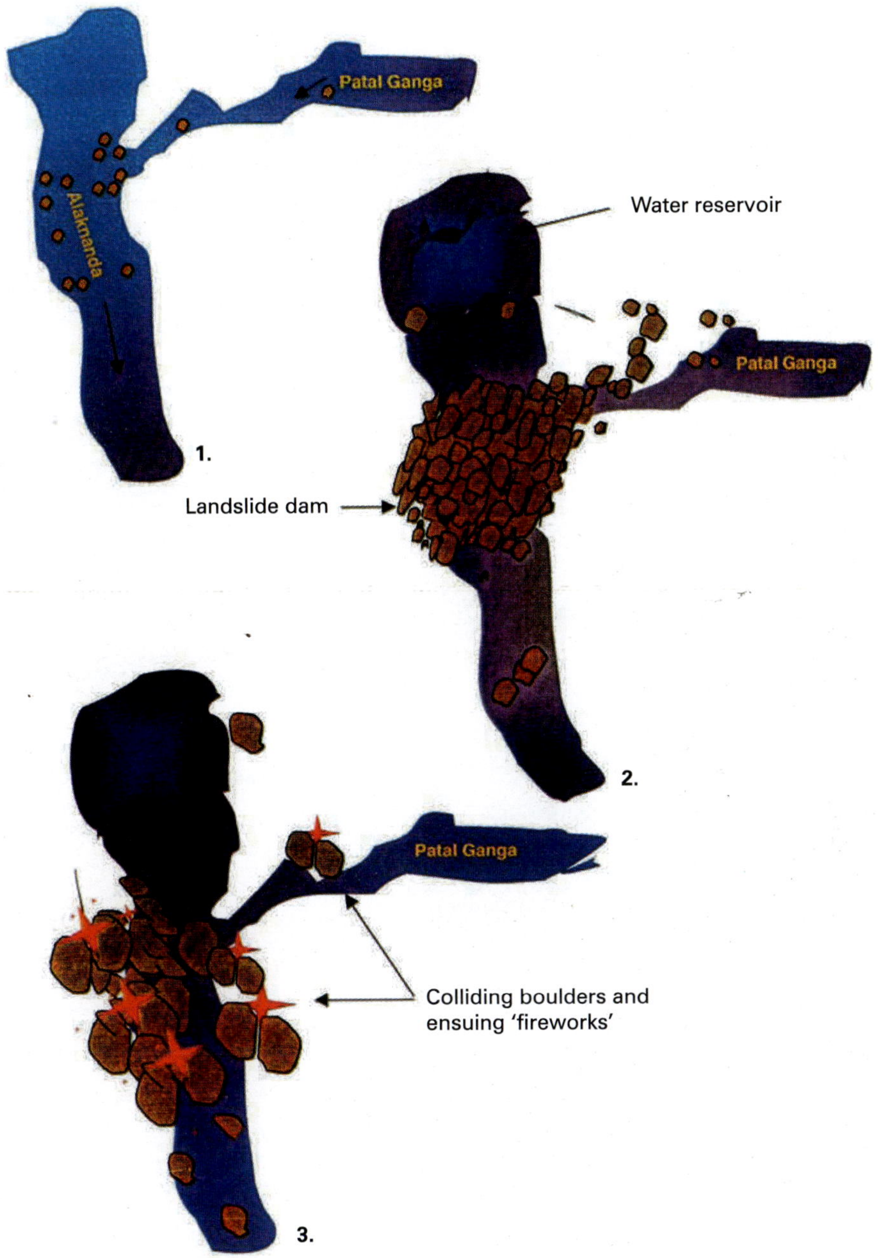

Plate 21 The Alaknanda tragedy triggered by the bursting landslide dam.

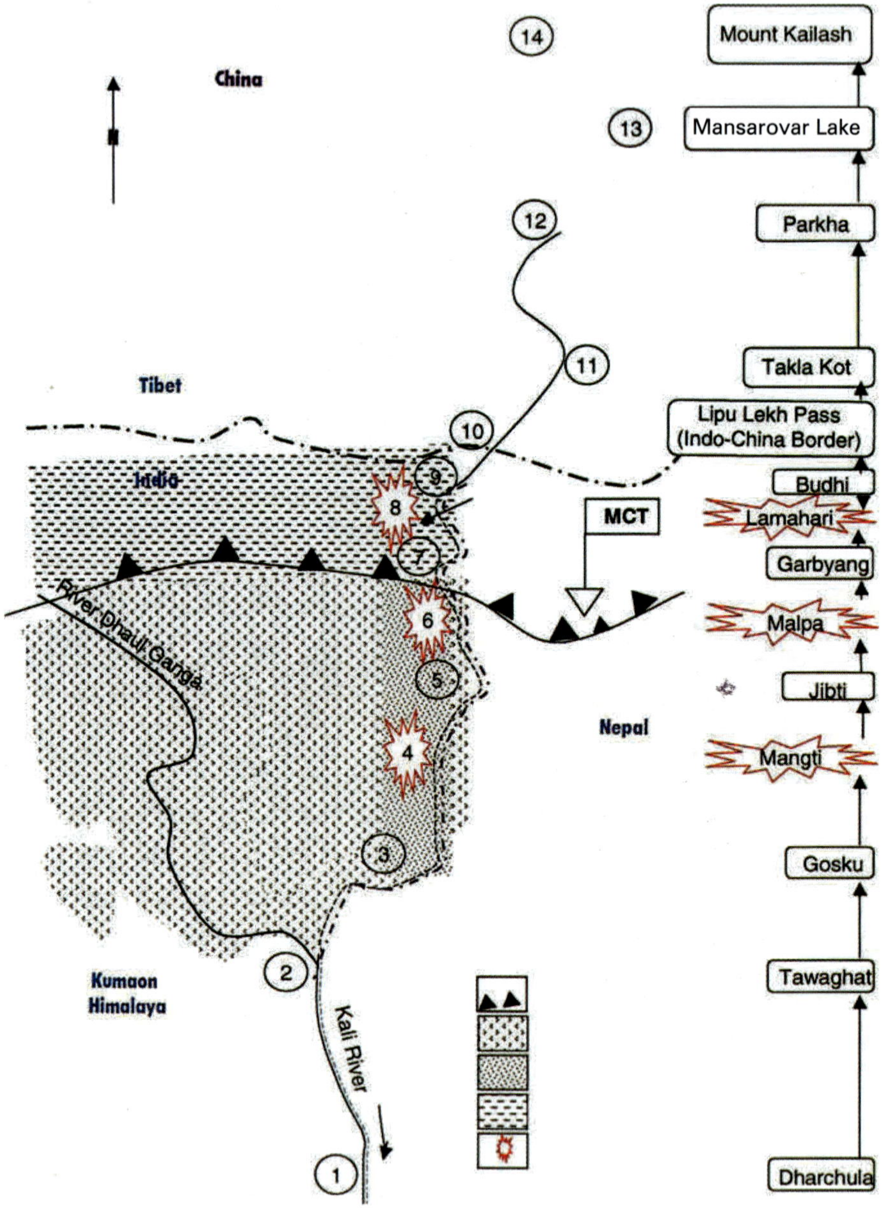

Plate 22 Location sketch of Malpa village and geology of the area.

It is an obvious but highly significant observation that when the load of debris in the river water grows huge and unmanageable, and when the continuous stimulus from rain is unavailable to maintain flow, the streams off-load debris, thereby raising bed levels, obstructing the path of flow and occasionally blocking it altogether, to create what is commonly known as a landslide dam.

The 40 m high landslide dam formed on the Rishiganga (also known as the Girthiganga) at Reni provides a notable example. The landslide blockade was made chiefly of augen-gneiss with subordinate bands of garnetiferous mica schist and quartzite. The lake thus formed became silted by May 1970, and the dam got breached during the July 1970 floods in Rishiganga. The left bank approach of the Reni Bridge, located 250 m upstream was also destroyed (Plate 8).

The Deadly Rainfall and other Causative Factors

Like most other tragedies, the Alaknanda tragedy was caused by an inter-play of several factors, varied and complex, inflicting staggering losses. To put it simply, the trinity of causative factors which unleashed Alaknanda tragedy were: (1) the unusual cloud-burst at Kunwari Khal, (2) the consequent formation of a landslide dam in the narrow constriction of river Patalganga and the simultaneous choking of the River Alaknanda, and (3) the eventual bursting of the landslide dam generating a formidable flood wave in the Alaknanda river.

The rainfall on the fateful day of the Alaknanda tragedy amounted to 212.8 mm, surpassing the highest one-day rainfall of 200 mm recorded at Joshimath on 28th September, 1924. This was only one quarter of the average annual rainfall of 990 mm at Joshimath (277 km; altitude 1905 m). Half the annual rainfall at Joshimath is known to occur during July (180 mm), August (190 mm), and September (110 mm).

Geology

The Alaknanda Valley could geologically be described as consisting of three major lithological units, viz: (1) the Dudhatoli Group, (2) Garhwal Group, and (3) Central Crystalline. The Dudhatoli Group refers to the zone between Devprayag and Koteshwar in the Bhagirathi Valley. It is constituted by moderately metamorphosed phyllites and quartzites. The north Almora thrust separates this group from the Garhwal group, exposed upto Vishnuprayag. The Garhwal Group consists primarily of quartzites, slates, schists, and carbonate rocks with meta-volcanic intrusions. The highest frequency of landslides in this group – such as those at Belakuchi, Nandprayag and Kaliasaur

– are sometimes attributed to a profusion of rock shearing and fracturing in the area. The northern zone extending from Vishnuprayag to Mana, and the neighborhood along the River Alaknanda and Dhauliganga, consists of schists, gneiss and granites of the Central Crystalline Group resting over the Garhwal Group.

Unfolding of the Alaknanda Tragedy

During the period 1st July to 19 July 1970, Rivers Alaknanda and Dhauliganga both reportedly carried a heavy load of debris contributed by the widespread landslides and primed by the rainfall of 126.5 mm, recorded at Joshimath. Then came something of a pause in the rainfall. From 8 am on the 19th July to 8 am on the 20th July, only 14.1 mm of rainfall occurred, and that must have surely taken some force out of the carrying power of the sediment-laden Alaknanda. That is probably how the huge amount of debris was off-loaded by the River Alaknanda and its tributaries, choking their courses, especially where the river constrictions were narrow, like the one at Patalganga (Plate 9), or that between Vishnuprayag and Pakhi.

This pause in rainfall was quickly followed in succession, by an all-time high rainfall of 212.8 mm between 2 pm on 20th July and 8 am on 21st July (i.e., in about 20 hours), surpassing the previous maxima of 200 mm.

The cloudburst prompted the River Patalganga to move a huge pile of debris driving the River Alaknanda to its right bank and severely choking it, see Plates 10, 11 and 12. Soon the Patalganga itself was blocked at its confluence point. Thereafter, the water level began to rise rapidly. The shifting of the Alaknanda to its right bank created a situation identical to the one found in the middle of a meander (Figure 1). A massive blockade also occurred on the Karamnasa Nala near Helang.

At 6 pm, immediately upstream of the Patalganga's confluence with the Alaknanda, the water level rose by 45.7 m within a matter of 45 minutes. The breach of the landslide dam occurred at about 6.45 pm, and the next one hour thereafter turned out to be perhaps the most devastating 3600 seconds of time in the known history of the area. Upon the breach of the landslide dam, the highest water level mark, seen between 256 km and 253 km, provided legible signatures of the abnormally high water impounding. The impounding led to deposition of huge thickness of debris on the road (Figure 2). One belief is also that when the Patalganga landslide dam burst, another dam was created by the debris so released in the already narrow channel of the Alaknanda. This, however, seems less likely because the bursting of a dam is usually accompanied with a tremendous amount of energy release, usually enough to flush out the debris.

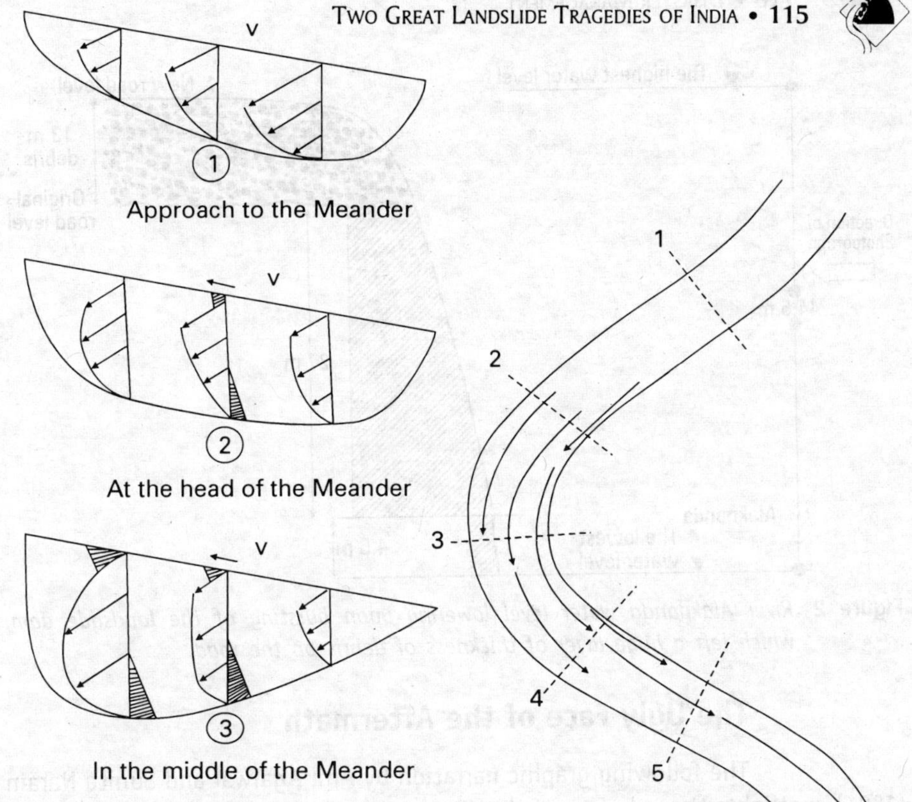

Figure 1 *A generalized diagram of expected distribution at a river meander. A situation similar to the position ③ above developed at the River Alaknanda when the debris discharged from the River Patalganga forced the flow of the Alaknanda to its right bank.*

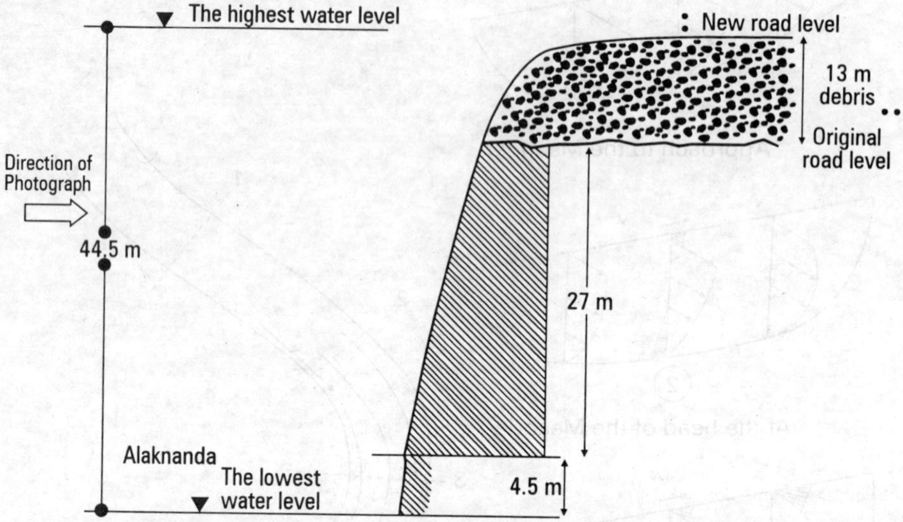

Figure 2 *River Alaknanda water level lowering upon bursting of the landslide dam, which left a huge layer of thickness of debris on the road.*

The Ugly Face of the Aftermath

The following graphic narration by Anil Agarwal and Sunita Narain (1991) spotlights the ugly face of the aftermath.

> "For the people of Alaknanda Valley, the night of July 20 was a nightmare. Nobody living along the river then thought they would see the next dawn. As the landslide dam broke, huge boulders were thrown far and wide. The roar of the deep gorge was not only deafening, but the sparks which rose when the huge boulders collided hitting each other made local people claim that the river was on fire that night."

The Damage Scenario

A brief account of damages is given below:

- A double tube English Bridge of span 40 m at Helang (261 km) was washed away (Plate 13). The Helang slide, located 13 km downstream of Joshimath, occurred on the bank of Karmansa Nala, upstream of its confluence with the River Alaknanda. The riverbed at this location had already silted up to the bridge deck level.
- The Langsi Bridge at 254 km was destroyed. The huge pile of debris brought down by the Langsi Nala is shown in Plate 14. A kilometer downstream, the Belakuchi slide (253 km), Plate 15, synergized with the flood wave in the Alaknanda valley unleashing what came to be known as the Belakuchi Tragedy

of July 1970. It is believed to have been initiated by the twin effect of severe slope erosion and movement along the foliation plane of a thick talc of schist band. As a result, the steel girder bridge was washed away. The Belakuchi Bazaar too was wiped out and so was the bus terminal; as many as 38 human lives were lost. It did not even spare the GREF Camp located at a height of 45.7 m from the riverbed. The other losses included 245 heavy vehicles including buses, 5 taxies and one army vehicle. It was here, between 253 km and 251 km, that the road completely disappeared and so also 15 culverts and 2 causeways.

- The Sial landslide located at 249 km, occupied an area that measured 300 m × 180 m. The area in this location settled between 5 m and 15 m.
- The Pakhi landslide occurred at 247 km. A fault zone may have been partly responsible for this slide.
- The road between Pipal Koti 245 km, and Birahi at 235 km underwent wide-spread subsidence as the trail of devastation travelled further downstream.
- A major blockade of the river upstream of the confluence of the Birahiganga with the River Alaknanda (Plate 16) in the Birahi gorge resulted in an afflux of 10 m–12 m. A 3 km-long stretch of road was wiped out, throwing a huge load of debris into the river. Nearly 500 m of the left bank, and the abatement of the Girder Bridge at Birahi, was washed away (Plate 17).
- The flood wave surged further down towards Chamoli (227 km), and a 300 m length of road was completely washed away.
- Downstream of Chamoli, right up to Srinagar, a total of 350 m of road length was breached, more than a kilometer of retaining walls and breast walls were lost, more than 2 km of drains were impaired, and as many as 40 culverts and 5 causeways were lost. After the flood receded, the road at Srinagar (131 km) was found to be under 1 to 2 m of silt.
- The Nandprayag landslide, around 216 km, in July 1970, measured 200 m in length and was 160 m wide. This slide was responsible for the repeated blockades of the road at this location (Plate 18).
- The River Alaknanda severely eroded its own left bank at Srinagar, and the situation at this location is known to be progressively worsening, year after year (Plate 19). At most places, the road formation was lost (Plate 20).

On the whole, on 20th of July 1970, within a short span of 45 minutes, a spate of landslides occurred; the most important among them were: the Helang, Belakuchi, Tangani, Sial, Pakhi, Birahi, and Nandprayag landslides. In the Dhauliganga Valley, besides breaching the landslide dam formed on the Rishiganga in 1968, the prehistoric Dhak landslide at Kuaripass was also reactivated due to serious erosion inflicted by the rather narrow and deep Dhak

Nala. A 300 m stretch of the Joshimath–Malari road sank by almost 40 m, more or less blocking the Dhauliganga, with the water level rising by 15 to 20 m. Bridges at Vinayak Cheti, Reni, Helang, Belakuchi, Langsi, and Birahi were among those destroyed, and the bridge abatements at Vishnuprayag and Birahi were also washed away. Large lengths of roads were either blocked by the landslides, or were completely breached; road subsidence was widespread. Severe erosion due to rivers in spate affected the lower parts of the slopes of valleys, provoking numerous new landslides, besides reactivating the old ones. The 1970 flood brought an estimated 9.1 million cubic meters of silt and rock into the Alaknanda, eroded largely from the catchments of the Patalganga and Garuganga.

Formation of a Landslide Dam in the Alaknanda Valley

A river blockade resulting from excessive sediment load may lead to the formation of a landslide dam, especially if provoked by a cloud-burst (Plate 21). The sediment load in the rivers is generated by the combined effect of erosion, creep, landslides, and water action. All the tributaries in the area do contribute, eventually loading the Alaknanda at their respective confluence points. The Dhaulagiri does so at Vishnuprayag, the Nandakini does so at Nandprayag, and the Pinder does so at Karnaprayag. The Mandakini does the same at Rudraprayag. And, in turn, the mighty Alaknanda itself does so at Devprayag.

To produce erosion, two requirements must be met. First, the rate of rainfall must be sufficient to produce run-off; second, the force of the moving run-off must outdo the resistance of soil material to erosion. *Inter alia,* it depends on the particle size (Figure 3). The creep due to gravitational forces also brings down the eroded material.

The alarming levels of slope denudation, creep, and landslides combine to increase water infiltration into the slopes, at least temporarily. The overland flow is thus restricted to about 10–12 per cent. The proportion of surface and subsurface water that feeds the river system depends upon the wide variations of geology and the tectonic and fluvial histories of the area, the complexity of denuded landforms, the diversity of vegetation, as well as the nuances of the man-made environment and its mindless growth.

The rivers and their tributaries are usually responsible for 85–90 per cent of the total sediment transport to the sea, the glaciers for about 7 per cent, the ground-water and waves for about 1–2 per cent, and the wind and volcanoes for less than 1 per cent. The total sediment transport by the Alaknanda may well cross the figure of 90 per cent.

In the Alaknanda Valley, the transportation by flowing water assumes dominance over the transportation by creep.

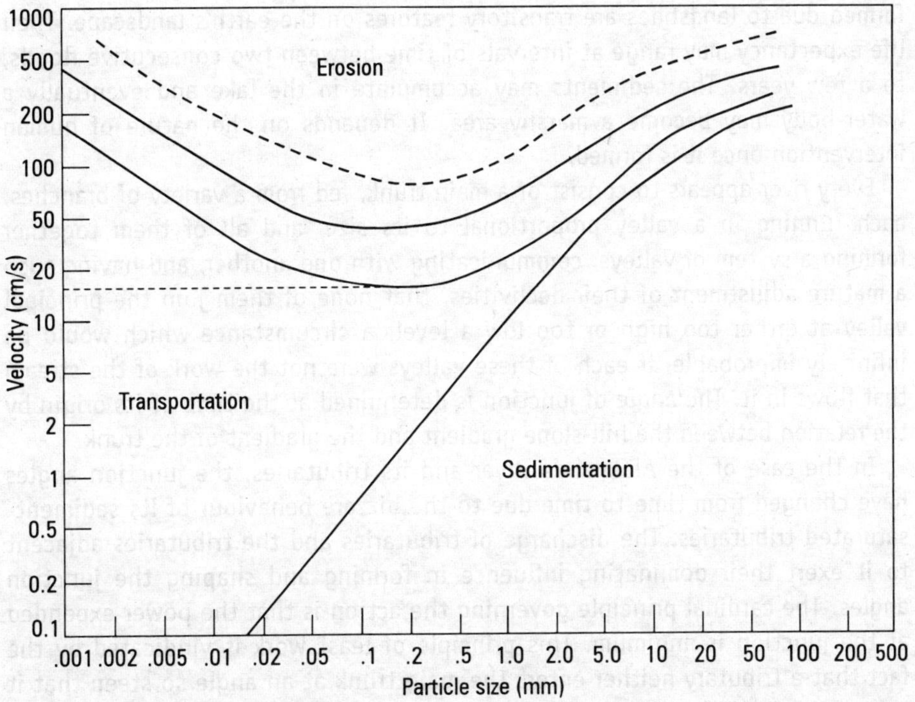

Figure 3 *Velocity and particle size largely decide the processes of erosion, transportation and sedimentation.*

Complexity of Sediment Transport

The rock masses are carried down partly by sliding, rolling, or saltating on or very near the bed, and partly in suspension by the fluid. The logical way to explain transportation would be to consider the force required to maintain movement at a given velocity. The product of velocity and force has the dimensions of power. In the case of bedload, the driving power is opposed by friction. Similarly, the power necessary to carry suspended loads is also a product of a force (weight) times velocity.

As the sediment concentration thickens, it is natural for the speed of particles to decrease relative to the fluid. A sort of traffic jam occurs in particle motion. When the bed debris is heterogeneous in size, the affinity of conglomeration turns high, resulting in the slowing down of sediment movement. The gradient of velocity near the bed may increase. If that is not sufficient to facilitate motion, sediments start settling down, the heavier stuff like boulders taking a lead. The fresh sediment load falls in queue, and a reservoir can be seen in the making. A single landslide can create a similar situation, if it is big enough to block a narrow constriction of a river. The lakes formed due to landslides are transitory features on the earth's landscape. Their life expectancy may range at intervals of time between two consecutive floods, to a few years. The sediments may accumulate in the lake and eventually a water-body may become a marshy area. It depends on the nature of human intervention once it is formed.

Every river appears to consist of a main trunk, fed from a variety of branches, each running in a valley proportional to its size, and all of them together forming a system of valleys, communicating with one another, and having such a mature adjustment of their declivities, that none of them join the principal valley at either too high or too low a level; a circumstance which would be infinitely improbable, if each of these valleys were not the work of the stream that flows in it. The angle of junction is determined at the time of its origin by the relation between the hill-slope gradient and the gradient of the trunk.

In the case of the Alaknanda River and its tributaries, the junction angles have changed from time to time due to the bizzare behaviour of its sediment-saturated tributaries. The discharge of tributaries and the tributaries adjacent to it exert their dominating influence in forming and shaping the junction angles. The cardinal principle governing the action is that the power expended at the junction is minimum. This principle of least work is vindicated by the fact that a tributary neither enters the main trunk at an angle so steep that it has to do any excessive digging, nor so gentle that the consequent loss of velocity unloads sediment at the junction.

Earlier Examples of Landslides Dams in the Alaknanda Valley

The month of August should indeed serve as an auto alarm, as most of the previous tragedies have occurred during this month or near about. A century ago, it was the month of August, to be precise, 26th August 1894, on which a massive landslide dam on the River Birahiganga burst to destroy the 4 km-long and 700 m-wide Gohna Tal, formed by a huge landslide in September 1893. The slide mass was made of limestone and dolomite; it covered an area of about 1.5 sq km. The breach of the Gohna Tal sent a flood wave down the River Alaknanda to destroy whatever came to oppose the flow, right up to the town of Srinagar. Decades later, when the Bhagirathi tragedy struck because of formidable flash floods, it was again month of August. A landslide dam was created by the Kanodia Gad Slide, which got breached on 10th August 1978. According to an eyewitness, the slide began to show up after the midnight of 6th August. It looked like a formidable torrent of mud which blocked the flow of the Bhagirathi for 14 hours forming a dam 35 m high and 45 m wide. The lake thus created was about 4 sq km in area. At another location, about 3 km

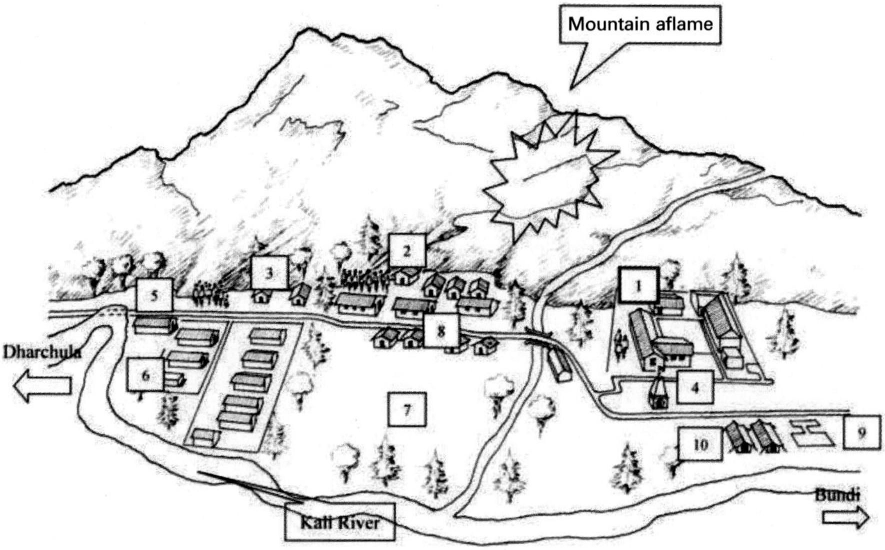

1 Eyewitness	6 KMVN Huts
2 Pilgrims and others with torches	7 Heap of debris
3 Footpath destroyed by avalanche	8 Civilian Huts
4 Temple	9 Helipad
5 PWD Huts	10 ITBP Camp

Plate 23 A bird's eye view of the Malpa rock avalanche of 18th August 1998, showing Malpa village in the foreground, which was instantly buried under a huge cover of rock debris.

Plate 24 Exposure at the hill face showing the scars of the rock avalanche.

Plate 25 No trace of the Malpa tragedy.

Plate 26 Before the dust storm (all green).

Plate 27 After the dust storm (all brown).

away, a second lake measuring 2.5 sq km in area, was formed when the River Kanodiagad was blocked at its confluence with the River Bhagirathi. The Lambagarh landslide tragedy on the Joshimath–Badrinath road, too, occurred on 2nd and 3rd of August 1997. The devastation following the breach of the Ukhimath landslide dam at the confluence of the River Madmaheshwar with Mandakini, also took place at Bhenti on 18th August 1998. The Ukhimath landslide travelled a distance of about 1.5 km, blocked the River Madhmaheshwar, and rode up the opposite slope by about quarter of a kilometer.

The Alaknanda tragedy stands out as the most spectacular example of an event, which struck on 20th of July 1970, i.e., a few days before the onset of August. Similarly, in September 1969, the huge Kaliasaur Landslide (147 km) blocked nearly three-fourths of the width of the River Alaknanda, and it remains a nightmare even today. It provides an outstanding example of how slides can deteriorate upon neglect.

Important Inferences

- Real drainage networks, such as the Alaknanda river system considered in this paper, are essentially random, rather than orderly. The issue of randomness versus determinism in a natural system is one of the most vexing issues in the philosophy of science.
- Several tributaries of the Alaknanda are known to result in subterranean flow through complex routes and are usually held responsible for triggering landslides. The River Patalganga seems to behave in this manner as its waters reportedly surface at Belakuchi. This observation needs further study.
- The mathematical concept of a regime canal or an ideal river, which neither scours nor fills its channel, was marked out for the canals of India (Leopold Maddock 1953; pp 43–48). Over a wide range of conditions, it was found that width, depth, velocity, suspended load, all increase as simple power functions of discharge. As a thumb rule, after heavy rain, the width of the river (in feet) increases approximate to the fourth root of discharge (cusecs); the mean depth increases approximate to the square root of discharge; and the velocity increases approximately by cube root of discharge. It follows therefore, that when the mean discharge of the river increases downstream, the channel depth and the mean current velocity all increase.

  It would however be patently wrong to freeze our understanding of the geomorphological processes on the strength of a meager pool of evidence, gathered through field observations for a period of time, which bears little sense on the geological time-scale.
- Downstream of the landslide dam, when a policeman in Belakuchi noticed that the river had no water, he signaled the alarm. About 400 pilgrims

to Badrinath, who happen to be at Bėlakuchi, escaped the wrath of nature probably because of the alarm. It certainly turned about to be a highly scientific indicator to sense danger for the one who had "an eye" to see.

Malpa Rock Avalanche of 18th August 1998

Introduction

A rock avalanche of formidable consequence struck the hamlets of village Malpa situated on the right bank of River Kali, in the district of Pithoragarh (India), of the Kumaon Himalaya, bordering Nepal (Plate 22). The village was traditionally inhabited by tribal people, engaged in trade with Tibet for generations. With the opening of the pilgrimage route to Mansarover and Mount Kailash in China, the Malpa village began to bustle with human activity as a base camp (Plate 23). The Kumaon Mandal Vikas Nigam (KMVN) established their cottages for pilgrims here. The Public Works Department (PWD), the Indo-Tibetan Border Police (ITBP), and the local tribal people also built their huts and buildings on the bank of the River Kali. On the fateful night between 17th and 18th of August 1998, a huge mass of rock got detached from the head region of the parent rock (Plate 24), broke into myriad pieces, and hurtled down the slope. The rock avalanche so generated killed 210 people including 60 pilgrims, eventually coming to rest. The heaps of debris created at the base of the slope were about 15 m high, and these included rock fragments as big as 5 m. The estimated velocity of the avalanche in some of its reaches was 30 m/sec. Before the arrival of the avalanche, the hill-slopes surrounding the Malpa village looked green, virtually without any visible signs of instability (Plate 25). When the rock avalanche came, the whole village was wiped out.

The Eyewitness Account

The Malpa rock avalanche is a rare and unique case record that provides an authentic narration of the rock avalanching phenomenon, unfolded before the eyes of an eyewitness (one of the three survivors of the Malpa Tragedy). Interestingly, the eyewitness was placed by providence at a vantage point. This happened because of his failure to find accommodation due to the unusually large crowd in the village Malpa on the fateful night between 17th and 18th of August. The mountain slopes were clearly visible to him, and so was the avalanche activity from the guest house where he stayed. A blow-by-blow account of the facts as seen by the eyewitness, and the other connected information gathered from him is given below:

Date	Time	Event
14 August 1998		A team from the Border Roads Organization (BRO) reached Budhi Village, 8 km north of Malpa
16 August		Bad weather and heavy, non-stop rain throughout the day, discouraged the team to travel to the Malpa Village for fear of landslides *en route,* especially at Lamahari.
17 August		Weather cleared. No rain at all, until 21.30 hrs.
	1600	The BRO team reached Lamahari and continued trekking towards Malpa.
	1830	Upon arrival, the BRO team witnessed an unusual crowd at the Malpa Village, and since no accommodation was available, it moved on to the PWD guest house located at a higher elevation on the slopes.
	2130	It started drizzling.
18 August	0025	He heard the thunderous sound of the rocks, hurtling down the slopes, towards the Malpa village.
	0030	The impact of falling rocks generated strong flashes of light in the head region of the slopes, simultaneously as the thunderous noise was heard.
	0030	Comprehending danger, many pilgrims came out of their huts with torches, in order to see what was happening.
	0035	A sandstorm swept the area with a formidable speed, while the loud noise of falling boulders continued. Some of the boulders also hit the roof and the wall of the room occupied by the eyewitness. The fury of the rock avalanche became increasingly clear.
		Falling boulders destroyed doors, windows, walls and even the bed inside the upper floor of the PWD guest house.
	0037	The worst that could be expected, happened: Malpa was wiped out. The falling of stones and boulders temporarily stopped, but the dust storm continued.
		In the time that followed, the help of the Indo-Tibetan Border Police (ITBP) was sought for post-disaster rescue and first aid was provided to those who were hurt.
	0430	The survivors moved on to the ITBP Camp upstream. Strangely enough, all the dogs of the area seem to have survived, but all the horses were killed, perhaps because dogs were let free but the horses were tethered.
	0525	The first message of the Malpa Tragedy was radioed from the ITBP Camp.

Date	Time	Event
	0545	The sporadic falling of chunks of rocks and boulders continued.
	0845	A doctor from the nearest place (Gunji) was instructed to go to Malpa.
		A temporary approach from the PWD track to the landslide area was constructed to enable the rescue operation.
		A temporary bridge was constructed over Malpa Nallah, at a new location, to provide an access to the site where remnants of KMVN huts, etc, lay buried under heaps of debris.
		The average height of the debris stacks was about 15 m
	1630	Medical aid arrived. Rescue work and the exhumation of the dead bodies commenced. A makeshift helipad was also construction by the BRO
19 August	1500	Boulders again started to fall, and the rescue operation was hampered.
20 August		A temporary walkway was constructed on the affected area. The loud sound of falling rocks continued throughout the night.
21 August		Landslide activity with sandstorms were seen to occasionally occur around 9 am.

On 4th August, the local people at Malpa had witnessed a flash of light on the mountain slopes, accompanied by a thunderous noise. These clear symptoms of instability were however ignored because the villagers of Malpa, over a period of years, believed that the thunderous sound and falling of boulders down the slope was a normal affair and not to be taken seriously.

Geology of the Area

The Kumaon Himalaya, confined by the borders of Nepal and Tibet, is one of the most fascinating segments of the Himalayan arc, and in many ways quite unique in its geological setting. The lithology of the area around Malpa represents an intricate system of folding, thrusting, metamorphism, and igneous action. The great Himalayan belt of Kumaon comprises of Pre-Cambrian metamorhpites of Central Crystalline with isolated, but sizeable, occurrences of some metasediments, gneisses, augen-gneisses, streaky gneisses, schists, granites, quartzites and amphibolites. The slopes are generally high and steep (60°–70°), and the rocks of the region are fractured. The Main Central Thrust (MCT) is known to pass through Budhi, only 8 km from Malpa.

The avalanching rock mass at Malpa consisted of fragments of massive quartzite inter-bedded with thin band of garnet-bearing sericite schist. The freshly-exposed rock faces of the hill show a series of parallel foliation planes and near-vertical joints, striking perpendicular to the foliation plane.

Some Exceptional Observations

The Malpa rock avalanche has left behind its indelible signature on the freshly-exposed hill surface, especially in the head region, where a huge rock mass was detached to feed the avalanche. The detached part of the rock mass generated: (a) a spectacular and huge rock avalanche, (b) a bright flashing light and sparks on the upper slopes, and (c) a dust storm.

The flashes of bright light and streaks of fire were described by the eyewitness as a curious dance of randomly jumping candles. This was accompanied with thunderous noise due to the sudden detachment of the rock masses and the subsequent destruction. The detached rock masses after splitting many times over, bifurcated into at least two unequal major parts, hurtling down the hill slopes along the most favorable of slope contours, eventually coming to rest at the foot of the slope.

The appearance of light flashing on the higher slopes could possibly be explained in terms of: (1) the release of static electricity upon the fracturing and tearing away of the rock masses in the detachment phase of the rock avalanche, and (2) the impact of falling rocks and their collision and attrition during motion, (Bhandari, 1999). The fact that similar observations are not easily to be found in the published case records may well be because eyewitness accounts of rock avalanches are very rare. Also, the non-reporting might have been because most avalanches do occur in the wetter manifestation of slopes, and it is the dry slopes that usually provide an ideal setting for frictional sparking and fire. The observations of sparking seen on the slopes at Malpa also finds support in the reports of forest fires during earthquakes, which are common in this very area. Rastogi (1999) has provided an added evidence of fire-like sparking seen recently in the Agastmuni area near Chamoli in India, where a cumulative fracture length of about 3 km was mapped, and the development of major fractures as big as 500 m long were seen to generate sparks.

The dust storm following the rock avalanche was perhaps generated due to sudden release of huge quantity of fine particles during the rock detachment, rock attrition and avalanche action. Its severity was physically felt by the eyewitness when the dust impacted on his face. This was established later on by the hard field evidence, seen as a veil of brown dust carpet that covered the green slopes all over a great distance, especially in the windward direction (Plates 26 and 27).

As far as known to the author, the closest parallel to the Malpa avalanche in the documented literature may be the old eyewitness account of the catastrophic Nanjiang rockslide–debris flow in Northern Sichuan, China, (Wang *et al.*, 1996). The speed of the slide was reportedly so high that the flying rock masses rammed into one another, acquiring a velocity of up to 40 m/sec. Like the Malpa rock avalanche, this case record describes the continuous thundering noise and flashes of light generated by the violent motion of rocks, which punctured the walls and roofs of buildings, the same way as they did at Malpa. Unlike the Malpa rock avalanche, however, the Nanjiang rockslide is reported to have occurred in the wetter manifestation of rock slopes. If all the available field information is analyzed against the backdrop of the observed facts and the previous experience in the area, one gets the feeling that the rock avalanche at Malpa occurred in the dry manifestation of the slope.

There are other reported examples of sparking and dust storms related to earthquake-triggered rock avalanches. For instance, Middlemiss (1910), while describing the great 1905 Kangra earthquake in India, did mention puffs of dust that arose at intervals of time. It was stated that: "Whenever a more than usually large slice of hillside collapsed, it was followed by a gigantic puff of dust simulating volcanic action". A similar observation is also made in the report of National Geophysical Research Institute of India following the Chamoli earthquake of 29th March 1999. It was observed that "Along with some of the landslides, rock and dust was seen falling even after one week".

Concluding Remarks

The Alaknanda and Malpa Tragedies, described in this paper, are both spectacular examples of large-scale devastation unleashed by landslides. Such tragedies often serve as live laboratories for the scientists to learn from the events and mitigate future disasters.

The paper highlights the importance of eyewitnesses in capturing facts which are otherwise lost, creating a permanent void in the ensuing scientific studies.

REFERENCES

Agarwal, A and Sunita Narain (1991) Floods, Flood Plains and Environmental Myths. *A Citizen's Report on State of India's Environment*. A publication of Centre for Science & Environment, New Delhi, p. 25.

Bhandari, R K (1999) Natural Disaster in Asia and the IDNDR Stimulus, *Proceedings of the XI Asian Regional Conference on Soil Mechanics and Geotechnical Engineering,* Seoul, 15–20 August.

Leopold, L B and Maddock, T (1953) *Hydraulic Geometry of Stream channels and some physiographic implications: US QS Professional Paper:* **252**: 57.

Middlemiss, C S (1910) The Kangra Earthquake of 4th April 1905, *Mem. Geol. Survey of India,* **37**.

Rastogi, B K (2000) Chamoli Earthquake of Magnitude 6.6 on March 29, 1999, *Journal – Geological Society of India*, **55**(5): 505–514.

Wang Sijing, Ding Enbao and Dy Yonglian (1996) An eyewitness of high speed rockslide—debris flow in Northern Sichuan, China. *Proc. of the Seventh International Symposium on Landslides,* Trondheim, 17–21 June, pp. 901-905.

DROUGHTS

D P Rao*

SUMMARY

Droughts and other natural disasters have devastating impacts on human life, economy and environment. Though it is not possible to fully repair the damage wrought by drought, efforts however, could be made to minimize the potential risks by developing early warning strategies, preparing both short- as well as long-term developmental plans to provide resilience, and to mobilize resources for providing immediate relief. Space technology offers immense potential in this endeavor by providing the necessary inputs for drought management strategies, namely early warning, assessment and monitoring, and combating or proofing. While satellite communication has significant potential for real-time dissemination of information and early warning, earth observation satellites enable the continuous monitoring of atmospheric as well as surface parameters contributing to these phenomena. India is one of the very few countries in the world using space technology for real-time monitoring of agricultural droughts under the "National Agricultural Drought Assessment and Monitoring System (NADAMS)" mission. The project involves the seasonal spatial monitoring of crop condition at sub-district level, by generating a Normalized Difference Vegetation Index (NDVI) from the National Oceanic and Atmospheric and Administration (NOAA) Advanced Very High Resolution Radiometer (AVHRR) data during the kharif season. The Wide Field Sensor (WiFS) data from the Indian Remote Sensing Satellites (IRS-1C and 1D) are being used for drought assessment and monitoring at *tehsil* and block/*mandal* level. Bi-weekly drought bulletins had been issued during the period 1989–91, and monthly crop and seasonal condition reports have been issued since 1992 for eleven agriculturally

* Ex-Director, National Remote Sensing Agency (Dept. of Space, Govt. of India), Balanagar, Hyderabad 500037, INDIA. Email: dprasada@hotmail.com

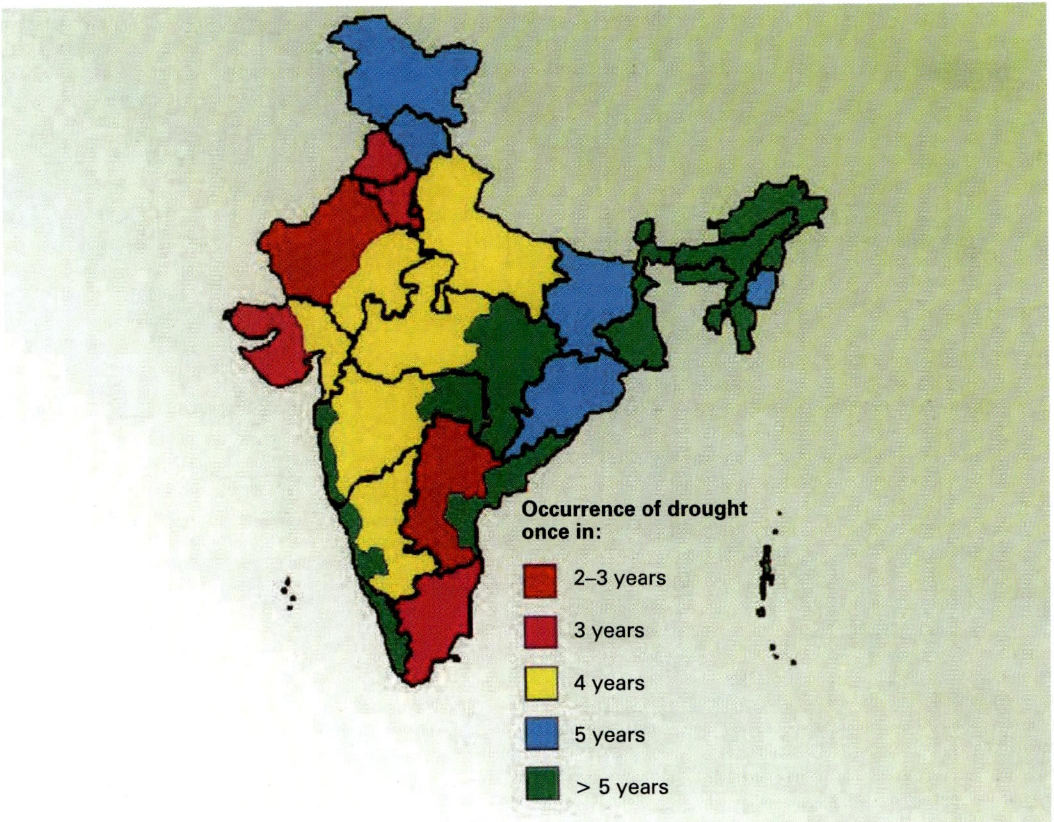

Plate 28 Frequency of occurrence of drought.

NATIONAL AGRICULTURAL DROUGHT ASSESSMENT AND MONITORING SYSTEM (NADAMS)

National reporting is done on administrative segmentation units:

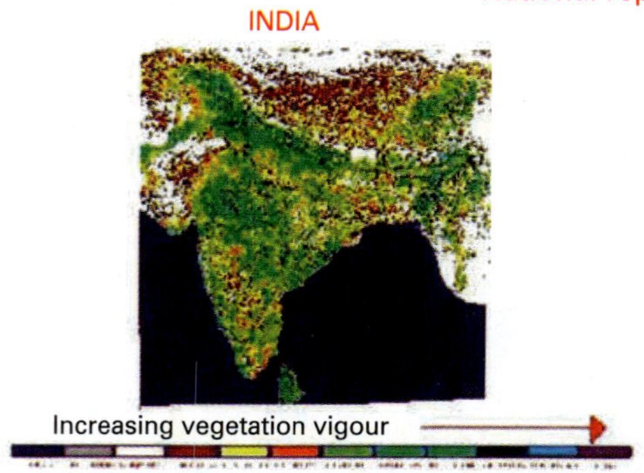

NOAA-AVHRR based NDVI: August 1999

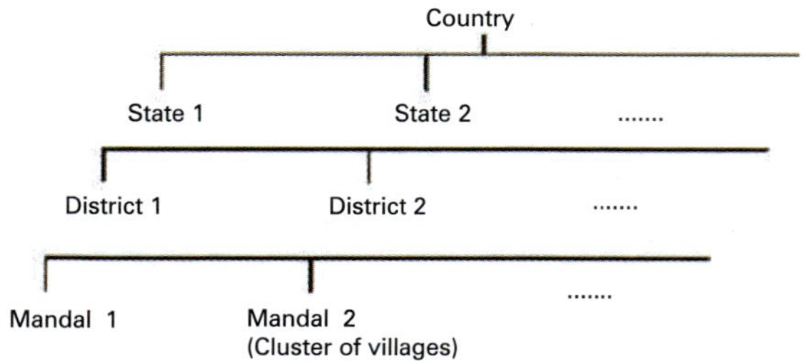

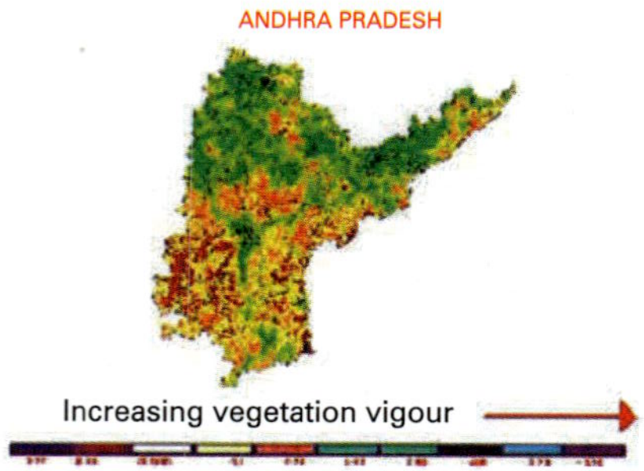

IRS-WiFS based NDVI: August 1999

Daily NDVI image is time-composited fortnightly and VI profile for crop season is compared with historic normal year

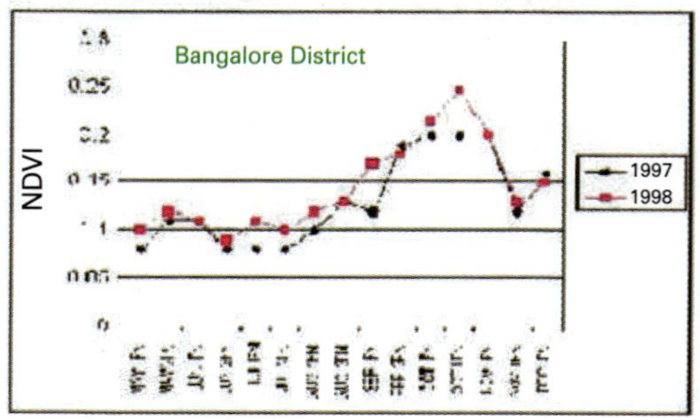

Plate 29 IRS-1D WiFS-based vegetation index maps of Andhra Pradesh and Cuddapah.

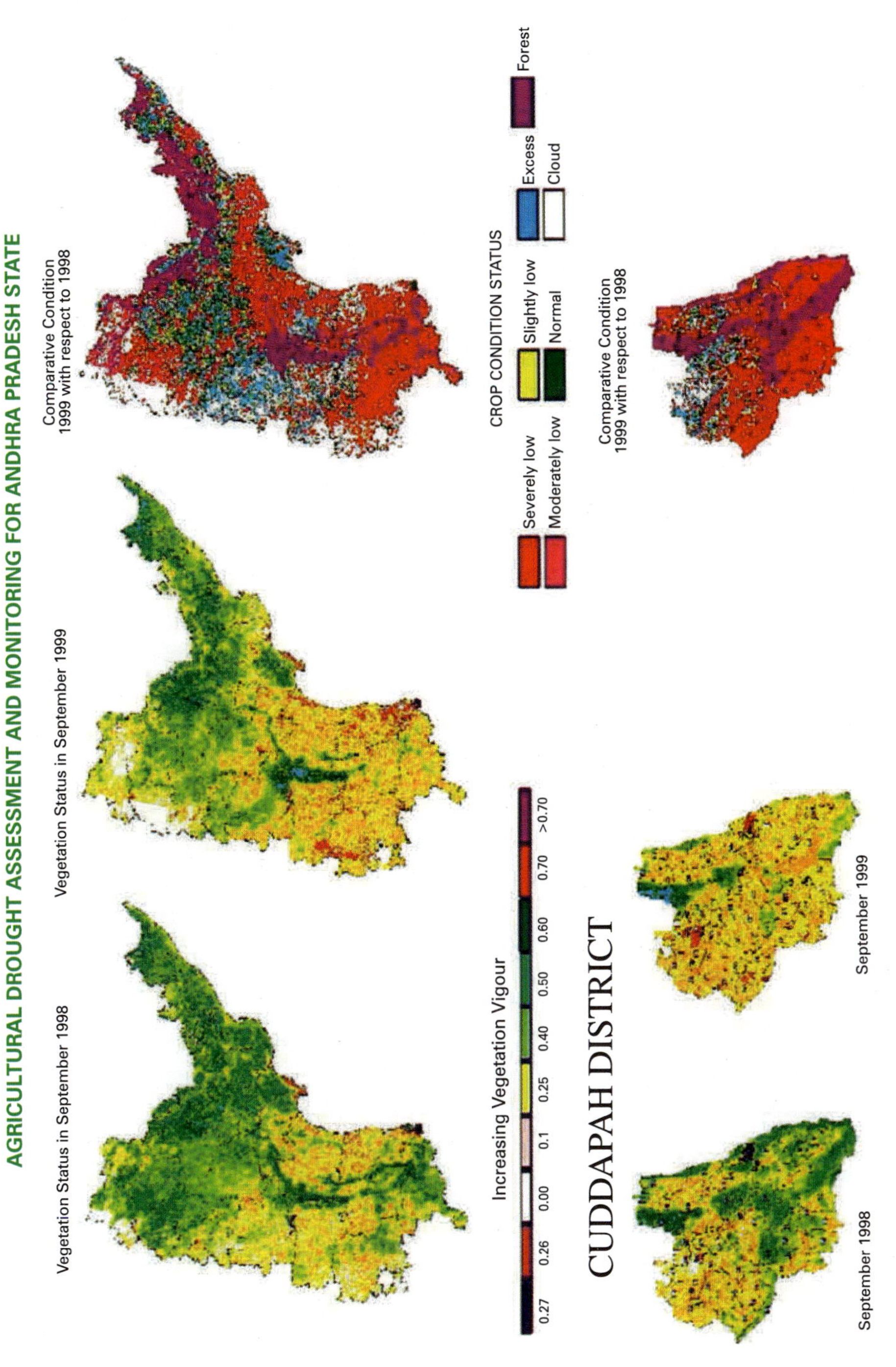

Plate 30 IRS-1D WiFS-based vegetation index maps of Andhra Pradesh and Cuddapah district.

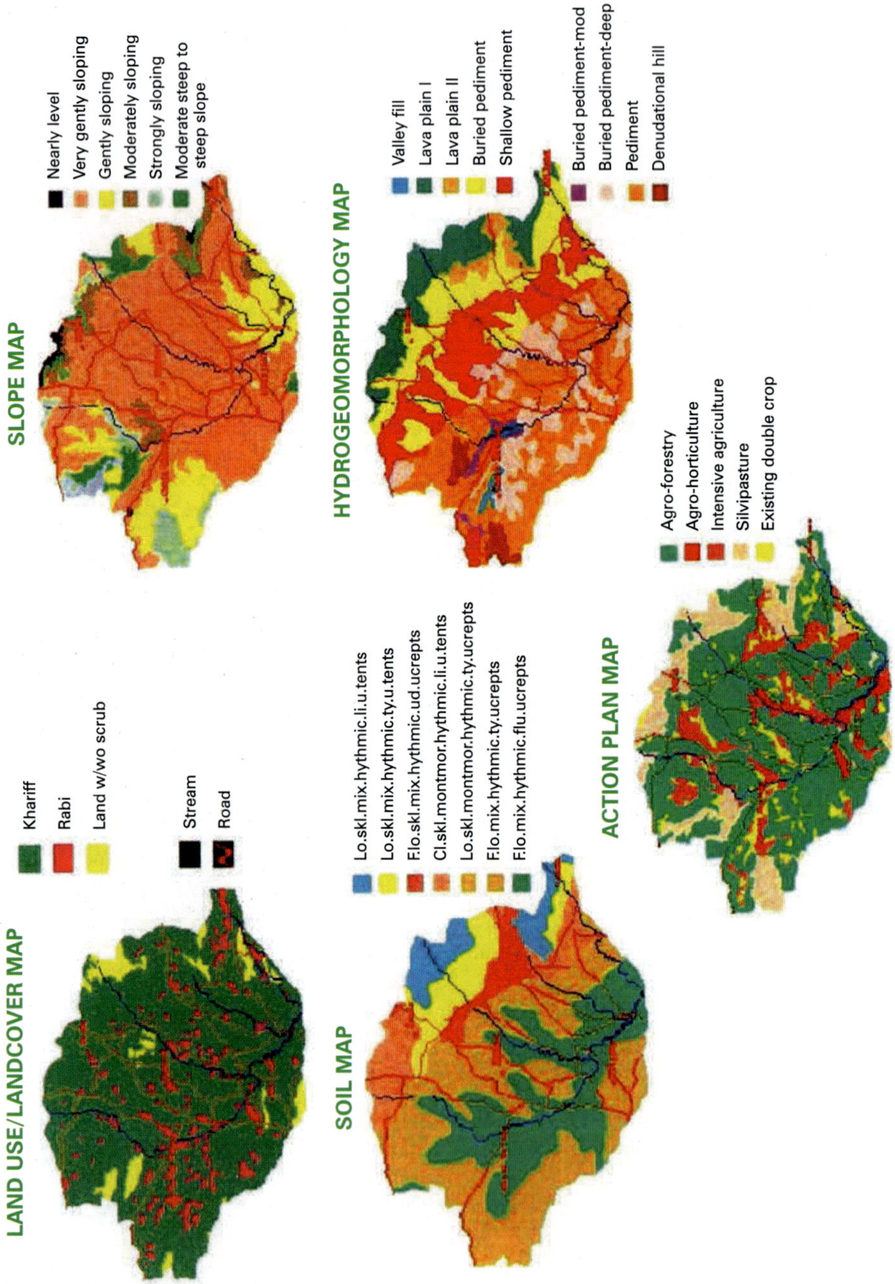

Plate 31 Integrated approach for drought combating—Case study over Upper Hatni Watershed, Jhabua district.

important and drought-vulnerable states in India. Based on the analysis of nearly 10 years of database, the establishment of quantitative relationships between vegetation index parameters and the yield of major crops in the districts is in progress. During the 9th five-year plan, a significant improvement in terms of detailed drought monitoring for the entire country at different administrative levels, and damage assessment using high spatial resolution Linear Imaging Self-scanning Sensor (LISS-II and III) data from IRS-1A, -1B, -1C, -1D and -P3, is envisaged. This article dwells upon various aspects of drought, the information requirement for its management, the role of space technology, the gap areas and the future scenario *vis-à-vis* likely developments in earth observation and communication technology.

Keywords: Drought management, remote sensing, drought preparedness, drought prediction, drought prevention, Normalized Difference Vegetation Index.

Introduction

Since 1900, major disasters have resulted in over 45 million deaths and have affected over 3.7 million people globally, resulting in severe damage and destruction of property and infrastructure (Paratesi, 1991). In practically all the developing countries, the agrarian communities are dependent on the vagaries of the monsoon and other climatic conditions. As a result, these countries are drought-prone. On an average, severe droughts occur once in every five years in most of the tropical countries; though often they occur in successive years, causing untold misery to human life and livestock. The Asian countries have been the largest victims of such disasters, accounting for over 60% of deaths and 85% of affected people globally. Drought (including famine) and flood alone are responsible for the largest number of deaths (over 53%), causing irreparable damage to property and ecology, and eventual economic loss (Jayaraman *et al.*, 1996).

Owing to the abnormalities in monsoon precipitation – in terms of both spatial and temporal distribution in India – drought continues to be a frequent phenomenon over many parts of the country. Out of the net sown area of 142 million hectares, about 68% of this area is reported to be vulnerable to drought conditions and about 50% of the drought-prone area is classified as severe, where the frequency of drought is regular. Coefficients of the variation of rainfall, derived from historical data, indicate the occurrence of drought once in 3–4 years in a major part of the country. The frequency of the occurrence of drought in India is shown in Plate 28. As evident from the figure, the major part of the country experiences drought once in every four years. The crisis brought about by this hazard directly hits the poorest and most deprived sections of our society, thereby destroying their life, economy, infrastructure, and environment. India has experienced four widespread major droughts after independence in the years: 1965–67, 1972–73, 1979–80 and 1986–87; and thereafter some part or other is invariably affected every year (Rao, 1998).

In 1979, India suffered from a widespread drought, which reduced the overall foodgrain production by as much as 20%. The drought of 1987, caused by the failure of the south-west monsoon over large parts of India, was one of the worst in this century. Only 14 out of 35 meteorological subdivisions within the country received normal/excess rainfall and 18 subdivisions received deficient rainfall. The overall deficiency during this monsoon period was (-) 19%. Such an order of deficiency was recorded only in two earlier years of this century, namely 1918 (-) 26% and 1972 (-) 25%. The prospects of the *rabi* crop in 1988 were threatened by the deficient rains in north-west India (Prakash, 1994). In the current decade also, large-scale intensive droughts have been observed in various parts of Bihar, Orissa, Andhra Pradesh, Karnataka, Rajasthan and Tamil Nadu very frequently.

The need to reduce the disasters and their impact received international prominence with the adoption of the United Nations General Assembly Resolution 42/169 in 1987 designated the 1990s as the International Decade for Natural Disaster Reduction (IDNDR). The decade's significance was further endorsed by the UN General Assembly Resolution 44/236 in December 1989, which proclaimed the decade and specified the policy and operational framework for pursuing the decade's goal.

Background

The frequent occurrence of widespread and severe droughts across the world emphasizes the need for systematic research on the causes as well as the impacts of drought, and the need for additional planning to help mitigate its worst possible impact. The major emphasis, thus, has been on the reconstruction of drought history and the computation of drought frequency. In addition, investigations on the first- second- and third-order impacts of droughts have also been carried out. Large-scale variations in the monsoon rainfall is caused by large-scale fluctuations in the atmospheric circulation during this period. Monsoon rainfall variations are determined by the behavior of the monsoon trough. The periodic shift of monsoon trough is the most important feature of the Indian monsoons. The rainfall areas also frequently shift in association with the changes in the monsoonal trough. A more southerly position of the trough is generally favorable for an active monsoon.

There is no universally acceptable definition of drought, although the word is associated with prolonged and abnormal rainfall deficiency. Rainfall is the single most important factor influencing the incidence of drought and practically all definitions use this variable singly or in combination with other meteorological elements (Bhalme and Mooley, 1980). For instance, Warwick (1975) has defined drought as: "A condition of moisture deficit sufficient to have an adverse effect on vegetation, animal and man over sizeable areas". Droughts are also frequently defined according to the disciplinary perspective. Subrahmaniyam (1967) has identified six types of drought, namely: meteorological, climatological, atmospheric, agricultural, hydrological and water management. Many others have included even economic or socio-economic factors as an essential factor in the determination of drought occurrence. Meteorological definitions are based on the degree of dryness or aridity and the duration of the dry period, whereas agricultural drought definitions link various characteristics of meteorological drought to agricultural impacts, focussing on precipitation shortages (Humphreys, 1931; Rosenberg, 1980), departures from normal (*World Book Encyclopaedia,* 1975), or numerous meteorological factors such as evapotranspiration (Laikhtman and Rusin, 1957). The definitions of hydrological drought, on the other hand, are concerned with the effects of dry spells on surface or subsurface hydrology, rather than a meteorological explanation of the event. For instance, Linsley *et al.* (1975) considered hydrological drought as: "A period during which streamflows are inadequate to supply established uses under a given water management system". The socio-economic definitions express features of the socio-economic effects of drought, and can also incorporate features of meteorological, agricultural and hydrological drought (Kifer and Steward, 1938).

Drought Management: The Conventional Approach

Though wholly avoiding damage due to drought does not seem feasible, effective drought management strategies, namely preparedness, prevention and relief, certainly help in minimizing its impact on agriculture and economy. Drought management involves the development of both short-term and long-term strategies. Short-term strategy includes early warning, monitoring and assessment of droughts; whereas long-term strategies aim at drought mitigation measures through proper irrigation scheduling, soil and water conservation, cropping pattern optimisation, etc. Rainfall prediction, weather forecasting, crop-yield prediction and identification of drought-prone areas comprise the preparedness phase. Included in the prevention phase are rainfall monitoring, hydrological monitoring and agricultural monitoring. The relief phase addresses food security, employment generation and a contingency crop plan apart from long-term relief measures, i.e., providing an assured drinking water supply, on-farm water harvesting, conservation and its effective utilization, alternate crop planning for delayed monsoons, raising drought-resistant crops, mid-season correction such as crop-thinning depending upon 'breaks in rains', and providing employment to the affected population through public works, providing cattle conservation programs and recovery strategies to enable farmers to resume their normal life when the next monsoon begins. These management activities can be grouped into the following three phases, namely the preparedness phase, prevention phase and relief phase.

Drought Preparedness

The activities in this phase include the identification of drought-prone areas and the prediction of drought to minimize its impact, which starts before the event. In India, the major emphasis is laid on drought-preparedness activities, such as identifying drought-prone areas which, generally, refers to the areas frequently affected by drought, and to carry out long-term drought mitigation activities that minimize the impact. The Irrigation Commission in 1972 suggested the following criteria for determining drought-proneness: (i) low rainfall regions, (ii) areas that receive irrigation support for less than 30 % of net sown area, and (iii) the frequency of famine and scarcity. Later, in 1976, the National Commission on Agriculture had made some suggestions for the selection of rainfall criteria. Areas with rainfall below 375 mm indicate extreme aridity, if the rainfall is between 375 mm and 750mm it is a semi-arid zone, between 750 mm and 1125 mm indicates a dry sub-humid zone, and the areas which receive above 1125 mm annual rainfall do not experience crop failure. The Task Force on the Drought-Prone Area Program (DPAP) and the Desert Development Program (DDP) evolved broad indicators, comprising of rainfall and irrigation with a relaxed irrigation yardstick to 40 per cent for areas receiving less than 750 mm rainfall and 30 per cent for areas with rainfall higher than 750 mm. The drought-prone areas identified, based on limited data on rainfall and irrigation details of the 1970s, were not revised so far in spite of an increase in the irrigated area, and changes in the land use during recent years. With frequent changes in the land use, irrigation development, cropping pattern and agricultural practices, it is, therefore, necessary to update the records of drought-prone areas with the enhanced understanding of the impacts of drought using satellite data.

Drought Prediction

Since drought results from deficiency of rainfall leading to soil moisture deficiency and, ultimately, to poor crop growth and yield, drought prediction involves the prediction of rainfall and crop yield.

Rainfall Predictions The prediction of drought is carried out mainly based on rainfall predictions. The rainfall predictions are of three kinds:

- *Long-range rainfall prediction* Since 1875, the seasonal rainfall forecast for the entire country as a whole has been provided by the India Meteorological Department, using statistical correlation utilising linear regression equations. Initially, the factors used were, essentially, the surface observation in and around India and later with the availability of upper air observations, upper air data were also included. In 1987, a parametric model which utilized 16 parameters was developed—some of these are global and others are regional in nature. The parameters are physically linked with the monsoon circulation. It has been observed that in a year when 60% or more parameters are unfavorable, the monsoon has invariably failed on such occasions. Since 1989, the forecasts have been fairly accurate due to the use of parametric and power regression models and dynamic stochastic transfer models. These forecasts are issued in two stages, the first tentative forecast is issued in mid April and final forecast in the last week of May. The forecast includes the seasonal total rainfall for the entire country.
- *Medium-range rainfall prediction* The National Center for Medium Range Weather Forecasting (NCMRWF) provides, in advance, the weather forecast at every 2.5° × 2.5° grid. Operationally, for 24 stations, 3-day forecasts are provided and the Agriculture Advisory Committee members from various Departments and Universities meet and take a decision on the necessary advice to be given to the farmers of these regions. The prediction is still at experimental stage and needs further refinement.
- *Short-range rainfall prediction* Based on the Indian National Satellite System (INSAT) data, supported by weather observations, qualitative predictions of weather valid for 24 to 72 hours is being issued daily.

Crop Yield Predictions Crop yield predictions are carried out based on rainfall and crop condition information received by the State Agriculture Department, and are compiled at national level by the Department of Agriculture and Co-operation, Ministry of Agriculture, Government of India.

Drought Prevention

This phase involves the continuous monitoring of the factors that cause drought, the assessment of drought impact and providing early warning to prevent or minimize the impact. Drought prevention measures include the following three activities.

Rainfall Monitoring

The India Meteorological Department (IMD), responsible for rainfall monitoring in the country, has meteorological observatories at each district headquarters and observes the weather information on a daily basis. In addition, the Agrometeorology Wing of the IMD regularly generates weekly aridity anomaly maps for the country. It identifies the areas with low moisture conditions at a coarse level, based on the water budgeting approach with limited data. Though the quality of observation is good, the rain gauge network is too coarse to account for variations of rainfall within the districts.

Hydrological Monitoring

Every day, water levels in all the medium and major reservoirs are monitored. The information is carefully analyzed by the concerned State Government Department to make policy decisions on the amount of water to be made available for irrigation, drinking water supply, etc. At the national level, the Central Water Commission regularly monitors the water levels of 47 major reservoirs. This information is passed on to the Crop Weather Watch Group, which alerts the Central Government in case of any abnormal conditions. The water level alone may not give the true picture of the availability of water. Periodic updating of the water level-capacity curve for each medium and major reservoir is required, due to regular siltation. All the medium and minor irrigation sources need to be monitored, as they are more susceptible to drought than the large reservoirs.

Agriculture Monitoring

The Directorate of Agriculture of every State has a well-established system to closely monitor sowing operations from village level to district and state levels. The system provides periodic information to Government on the area under various types of crops, the progress of agricultural operations, and the health and growth of the crops sown. The Crop Weather Watch Group at the Department of Agriculture and Co-operation analyzes the information at the national level and makes recommendations to the State and Central Governments for initiating a contingency crop planning program if necessary.

Drought Relief

This phase comprises of the adoption of short- and long-term relief measures. The State Governments are primarily responsible for the management of all natural disasters; the Central Government supplements the efforts of the states. The district level administration is responsible for the operational management. Drought impact assessment at the district-level involves the generation of a drought memorandum, indicating the losses and types of relief measures being adopted. The relief measures include:

- *Food security* The procurement, storage and distribution of food grains from the food surplus areas to the deficit areas, and the distribution of essentials through the public distribution system.
- *Employment generation* The preparation of employment generation schemes, and their technical evaluation for providing jobs to the drought-affected population.
- *Contingency crop plan* An alternate crop plan with sufficient infrastructure available to stabilize farm-level production. The contingency crop planning strategy involves three major approaches.
- *Crop life-saving techniques* It comprises of suitable cultural practices to conserve soil moisture and suppress evaporation; an alternate crop strategy, to improve crop production and social security schemes. Other measures include providing drinking water, establishing cattle camps, and waiving loans and providing monetary relief. Most of the above-mentioned drought impact minimization activities are carried out with inadequate data on current land use, land cover and other hydrological conditions.

The Management Issues

There are a few important issues related to drought management (Sharma, 1999). First, the Government seems to be willing to develop long-term strategies for drought management as long as the memories of the event are fresh. It does not recognize that disaster mitigation may be carried out at a minimal cost, simply by adjusting the on-going developmental programs. Second, owing to the lack of a clear concept of the broad range of issues that drought covers, very little emphasis is placed on institutional strengthening, and more focus is given to stockpiling and developing emergency response plans. Third, there is also a problem of inadequacy in developmental planning, with planners often lacking the essential technical information and expertise to integrate disaster risk into their models. Fourth, in the planning process, there is a predominance of centralized planning systems, which do not give sufficient importance to the views and concerns of regional and local governments. Consequently, the desired results are not achieved. Furthermore, due to attitudinal barriers that cut a cross a variety of stakeholders, the Government is sometimes reluctant to use the expertise of NGOs and the private sector, which essentially precludes people's active participation in drought management. Finally, preventive measures like watershed-based soil and water conservation, though advocated very often, are not implemented in letter and spirit due to various reasons. In cases of prolonged drought, it may result in the total loss of land productivity.

National Approach

Drought management in India was evolved as a national approach and has undergone significant changes in the light of the status of resources and technological developments. The famine codes, based on recommendations of Indian Famine Commission (1880), constituted a milestone in the history of drought management in India. This was the first time that a systematic approach was used. In the 1950s, famine codes were replaced by scarcity relief, with the objective of preventing starvation deaths. In the mid-60s, drought response mechanisms were built up to ensure physical and economic access to food. Since 1970, drought management is being carried out by providing relief and by adopting drought impact minimization through short- and long-term drought mitigation measures.

The primary responsibility of drought management in India rests with the State Governments. The Central Government supplements the efforts of the State Governments in dealing with disaster situations, and also provides the major part of the financial resources for disaster response. The Natural Disaster Management (NDM) Division within the Department of Agriculture and Co-operation, Ministry of Agriculture, and the National Centre for Disaster Management (NCDM) Division within the Indian Institute of Public Administration, are the nodal agencies for natural disaster management. There is an institutional arrangement at the national, state, district and sub-district levels to deal with emergency situations. A National Contingency Action Plan exists for ensuring emergency assistance in the wake of natural disasters at the national, state and district levels. State Governments have their Relief Manuals/ Codes which lay down the procedures and defines powers for emergency management and provision of relief.

The institutional arrangements at the national level consist of a Cabinet Committee on Natural Disaster Management and Crisis Management, presided over by the Prime Minister. Disaster relief coordination is effected by a Central Relief Commissioner in the Ministry of Agriculture and Cooperation. Each State Government has a Relief Commissioner and State Cabinet level coordination committee. At the district level, the District Collector presides over the relief committee, which consists of people's representatives. A Calamity Relief Fund (CRF) is allocated to each state on an annual basis, 75% of which is contributed by the Central Government. The quantum of the CRF is determined by independent Finance Commissions once in five years.

The Management Issues

A sound drought management strategy essentially involves long-term planning for establishing a systematic assessment, monitoring and predictive capability and adequate measures to combat drought. As mentioned earlier, as a consequence of the incidence of drought, agricultural crops, animals and humans are affected. Supplemental/life-saving irrigation to affected crops, and for drinking, domestic uses and for animals additional water is required, for which ground water is the only source. Long-term planning and provision for adequate funds for ground water exploration are pre-requisites. Water for both drinking as well as for irrigation is the most crucial element, which needs to be provided to drought-affected areas. Water harvesting both on-farm as well as in the rural and urban agglomeration needs to be planned on a long-term basis. The identification of sites ideal for ground water recharge, construction of structures though planned very often but their implementation, which requires human resources, funds and technical know-how, is often lacking. Watershed-based planning and management of water resources needs to be effectively implemented. Further, timely information on the sites for ground water exploration is required which is, in general, not available at the village level. Potential ground water zones need to be identified and the information may be made available to the users.

For storage of soil moisture during droughts several techniques, namely, delayed sowing, broad base and furrow system of cultivation, mulching, thinning of crops, etc, have been developed. In spite of the development of these improved crop management practices for the late onset of the monsoon or during long dry spells, acceptance by the farmer has not yet been fully realized.

Due to logistics and other procedural reasons, relief material especially food is generally not timely available to the affected people. There is a need to streamline the flow of relief material to the beneficiaries.

The Role of Space Technology

By virtue of a synoptic and repetitive coverage at regular intervals, space-borne measurements offer immense potential in generating information on various terrain and weather parameters required for drought management. Space technology plays a key role in all the phases of drought management—namely preparedness, prevention and relief. Remote sensing data provide major inputs to all the three types of rainfall predictions. Parameters such as global and regional atmosphere, land and ocean parameters – namely, temperature, pressure, wind, snow, *El Niño*, etc – required for long-term rainfall prediction are estimated or observed using remote sensing data from geo-stationary and polar-orbiting weather satellites, such as the INSAT and NOAA series of satellites. For medium-range weather prediction, the National Centre for Medium Range Weather Forecast (NCMRWF) uses satellite-based Sea Surface Temperature (SST), Normalized Difference Vegetation Index (NDVI), snow cover area and depth, surface temperature, altitude, roughness, soil moisture at surface level and TOVS and Radiosonde data on water vapour, pressure and temperature as vertical profile data. In short-range rainfall prediction also, INSAT-based visible and thermal data are being used.

Early Warning of Drought

The early warning of drought is useful for on-farm operations and to arrive at an optimal local water utilization pattern. Rainfall anomalies as observed from geostationary/meteorological satellites are being used for early warning of drought, which is yet to be fully operationalized. Studies have indicated that certain large-scale meteorological patterns are associated with the failure of the summer south-west monsoon, which is the main cause of droughts in the Indian subcontinent (Rao, 1988). The factors that can provide an early indication of possible droughts include upper air winds over India, the development of hot, low-pressure areas over Southern Asia, and the *El Niño*/Southern Oscillation phenomena in the Pacific Ocean. Other factors that can be observed by satellite and which are related to rainfall patterns are sea surface temperatures, snow cover, cloud patterns, wind velocity and humidity profiles. Geostationary and polar-orbiting, low earth-orbit satellites provide an excellent means of deriving most of the information on regional and global scales. Research is under way to develop more accurate models involving the atmospheric, marine and land factors for forecasting monsoon development and, even more importantly, the break in monsoon that results in droughts. These efforts may possibly lead to timely drought forecasting.

Assessment and Monitoring

The monitoring and assessment of droughts are required for initiating corrective measures at appropriate times, to minimize the reduction of agricultural productivity in drought-prone areas. This monitoring and assessment of droughts also provide objective information on the prevalence, severity level and persistence of drought conditions in a time-effective manner, which will help resource managers in optimally allocating scarce resources. The satellite-derived vegetation index (VI) which is sensitive to vegetation stress is now being used continuously to monitor drought conditions on a real-time basis, often helping the decision-makers in initiating strategies for recovery by changing cropping patterns and practices. The use of meteorological satellite data to assess the spatial and temporal inadequacies of rainfall at critical crop stages and the subsequent assessment of crop status/condition based on VI anomalies provide an excellent drought monitoring mechanism. However, space-borne measurements have to be integrated with the computed aridity anomaly, based on field measurements of rainfall and crop calendars, to bring out the real-time drought conditions of a region.

Sponsored by the Department of Agriculture and Co-operation, Ministry of Agriculture, Government of India, the **National Agricultural Drought Assessment and Monitoring System (NADAMS)** was established in the Department of Space in 1989, with support from the State and other Central Government departments. The approach relies on the fact that the vegetation strongly absorbs the visible light of electromagnetic energy and is related to absorbed, photosynthetically active radiation and biomass and absorbs little or no light and reflects more in the infrared region and is related with canopy geometry and the health of the crops. The effect of drought is to cause recognizable structural changes in vegetation, which in turn cause changes in the visible and infrared spectral reflectance. The structural changes include morphological changes such as leaf appearance (colour and shape) and stature of the plants and phenology. The vegetation index (VI) is the combination of the visible and infrared spectral response to enhance the vegetation condition. Repetitive satellite coverage provides us with the capability to monitor spectral changes resulting from the drought. The temporal pattern of VI is diagnostic of vegetation conditions. Any lowering of VI values reflects moisture stress in vegetation, resulting from prolonged rainfall deficiency. Such a decrease could also be caused by other stresses such as pest/disease attack, nutrient deficiency or geochemical effects. The seasonal VI profile, thus, reflects the vegetation dynamics and condition. A comparison of the VI profile of the reporting year and a previous normal agricultural year provides the assessment of drought impact in the scale of previous agricultural scenario.

In this project, the NOAA-AVHRR (1.1 km)-based biweekly Vegetation Index (Plate 29) is used daily for providing periodic information on crop conditions at district level (Jeyaseelan and Rao, 1997). It provides early warnings on expected yields from major crops in the district by the end of August, the end of

September, the end of October, based on the relationship between vegetation index growth profile obtained and the major crop, using historic data. Extensive ground data were collected for methodology development and validation. Ground truth campaigns were conducted with the cooperation of field officers of State Government departments. The calibration of VI datasets with ground data helped in establishing the VI and yield relationship and provides an early warning of vegetation development with the current status of rainfall. The results indicated that even when VI data is averaged over the district (an administrative unit) it reflects the major crop dynamics and conditions in the district, which is driven by the rainfall cycle during the *kharif* season (Plate 29). The results of the study are reported in the form of biweekly bulletins. Initially, biweekly drought bulletins for 246 drought-prone districts of the country were issued to the concerned user agencies at all levels for necessary action. Based on user feedback, the second phase of NADAMS was launched in 1992, which includes a detailed monthly drought assessment in terms of spatial variability and impact on crop/fodder production.

With the operationalization of the Indian Remote Sensing Satellite (IRS-1C) Wide Field Sensor (WiFS) and IRS-P3 WiFS and SWIR bands, our in-season agricultural drought monitoring capability has been further improved (Plate 30). Plate 30 shows the vegetation condition in Andhra Pradesh (AP) during September 1998 and 1999, derived from IRS- WiFS data. A comparison of the vegetation condition during 1999 as compared to 1998 is portrayed in the third image (Plates 29 & 30). The images in the lower half of Plate 30 exhibit vegetation conditions in the Cuddapah district of AP during September 1998 and 1999, derived from IRS WiFS data, and the relative vegetation condition during 1999, with reference to that of 1998.

Drought Combating/Proofing

Drought proofing calls for an integrated approach, taking into account the multi-dimensional inter-linkages between various natural resources and the environment on one hand, and the mutual interdependencies of natural resources on the other. The satellite remote sensing-based Integrated Mission for Sustainable Development (IMSD) is a unique Indian experience to evolve locale-specific action plans for the development of land and water resources towards combating droughts against the backdrop of the socio-economic conditions of a micro-watershed. The integrated approach of utilizing the existing conventional data with satellite remote sensing data assumes greater significance in the context of developing operational methodologies for basic resource mapping and management, to formulate long-term droughts mitigation measures (Rao, 1993). Information on natural resources derived from space-borne multi-spectral data are integrated with the terrain's slope and the socio-economic conditions in a Geographic Information System (GIS) environment to generate a locale-specific action plan (Rao, 1998). Here, information on hydrogeomorphology, soils, land use/land cover has been integrated with the terrain's slope and socio-economic data to generate the action plan. A case study following such an approach in part of Jhabua district is presented in Plate 31. Included in the action plan are various soil conservation measures like contour bunding, terracing gully plugs, water harvesting structures such as check dams, sub-surface dikes, etc. In addition, alternate land use, i.e., agro-horticulture, agro-forestry and silviculture are also suggested in order to optimally utilize the land while, at the same time, maintaining harmony with the environment.

Conclusions

In spite of rapid economic developments, and scientific and technological advancements, droughts are more frequent and are causing more damage, ultimately affecting social and economic developments. Space technology offers immense potential for early warning of and addressing most of the issues related to drought management on a near-real time basis. India has developed a space-based strategy for all the three phases of drought management, namely the preparedness phase, the prevention phase and the relief phase. Plans are afoot for a comprehensive assessment of drought by accounting for variations in rainfall, surface water availability and soil moisture, in addition to vegetation condition and communication through NICNET, for two-way data transfer to districts. Apart from general crop condition assessment, crop-specific damage assessment is also contemplated using high spatial resolution satellite data. Future earth observation missions with improved spatial spectral and temporal resolution may enable the realization of our goals especially in the relief phase of drought management, by way of enabling the refinement of the locale-specific action plan for land and water resources development aimed at providing long-term relief.

REFERENCES

Bhalme, H N and Mooley, D A (1980) Large-scale droughts/floods and monsoon circulation, *Monthly Weather Review,* American Meteorological Society, **109**(8): 1197–1211.

Humphreys, W J (1931) How droughts occur, *Bull. American Meteorological Society,* **12:** 18–22.

Jeyaseelan, A T and Rao, D P (1997) Satellite Contribution to Natural Disaster Management in India, *Conference on Geomatics in Disaster Management, September 18–20, 1997,* Dehradun, India.

Jeyaseelan, A T and Thiruvengadachari, S (1986a) Current drought monitoring system in Andhra Pradesh state. *Report No. IRS.UP.NRSA-DRM-JR-03-86,* NRSA, Hyderabad, India.

Jeyaseelan, A T and Thiruvengadachari, S (1986b) Recent drought history of Andhra Pradesh state, *Report No. IRS.UP.NRSA-DRM-JR-04-86,* NRSA, Hyderabad, India.

Jayaraman, V, Chandrasekhar, M G and Rao, U R (1996) Managing the natural disasters from space technology inputs. *Proc. 47th International Astronautical Congress. Oct. 7–11, 1996,* Beijing, China.

Kifer, R S and Steward, H L (1938) Farming hazards in the drought area, *Monograph XVI,* Works Progress Administration, Washington, D C.

Laikhtman, D L and Rusin, N P (1957) On the meteorological criterion of drought (Russian), *Glavnaia Geofizicheskaia Observatoriia,Trudy,* **69**(1957): 65–70.

Linsley, R K, Kohler, M A, and Paulhus, J I H (1975) *Hydrology for Engineers* (Second edition), Mc Graw-Hill, Kogukusha, Tokyo.

Paratesi, S R G D (1992) *Hazards and Disasters: Concepts and Challenges, Remote Sensing for hazard monitoring and disaster assessment,* Barett, E C *et al.*, Beach Science Publishers, Paris.

Prakash, I (1994) *Disaster Management,* Rashtra Prahari Publishers, Sahibabad, Ghaziabad, India, p. 91.

Rao, U R (1993) *Space and Environment, Proc. IAF Congress,* Austria, Graz.

Rao, U R (1998) *Space and drought management. Current session, 39th IAF Congress,* Bangalore, Special publication, ISRO-SP44-**89**(5–11).

Rosenberg, N J (Ed) (1980) *Drought in the Great Plains: Research on Impacts and Strategies,* Water Resources Publication, Littleton Co.

Sharma, V K (1999) Status of preparedness planning in India for disaster mitigation, *Shelter,* Special issue—World Disaster Mitigation Day, October 13, 1999, A HUDCO-HSMI publication, pp. 13–15.

Sinha, D K (1991) Coping with natural disasters: An integrated approach. In: D K Sinha (Ed) *Natural Disaster Reduction for Nineties: Perspectives, Aspects and Strategies,* International Journal Services, Calcutta, p. 23.

Subrahmaniyam, V P (1967) Incidence and spread of continental drought, *WMO/IIID report,* No **2**, Geneva.

Warwick, R A (1975) *Drought Hazards in the United States: A Research Assessment,* Boulder, Colorado, University of Colorado, Institute of Behavioural Science, Monograph No.NSF/RA/E-75/004, p. 199.

World Book Encyclopaedia (1975) Field Enterprises Educational Corporation, Chicago.

Index